SETTING OUT

A GUIDE FOR
SITE ENGINEERS

S.G. Brighty

Second Edition

Revised by

D.M. Stirling BSc

Lecturer in Surveying & Photogrammetry
The City University, London

BSP PROFESSIONAL BOOKS

OXFORD LONDON EDINBURGH

BOSTON PALO ALTO MELBOURNE

First published in Great Britain 1975 by
Crosby Lockwood Staples
Reissued 1981 (with revisions) by Granada
Publishing Limited
Reprinted 1982
Second edition by BSP Professional Books
1989

British Library
Cataloguing in Publication Data

Brighty, S.G.
    Setting Out. — 2nd ed.
    1. Construction. Sites. Setting out:
    Manuals
    I. Title   II. Stirling, D.M.
    624

    ISBN 0−632−02039−3

BSP Professional Books
A division of Blackwell Scientific
    Publications Ltd
Editorial Offices:
Osney Mead, Oxford OX2 0EL
    (Orders: Tel. 0865 240201)
8 John Street, London WCIN 2ES
23 Ainslie Place, Edinburgh EH3 6AJ
3 Cambridge Center, Suite 208, Cambridge,
    MA 02142, USA
667 Lytton Avenue, Palo Alto, California
    94301, USA
107 Barry Street, Carlton, Victoria 3053,
    Australia

Set by Setrite Typesetters Ltd
Printed and bound in Great Britain by
Mackays of Chatham PLC, Chatham, Kent

Contents

Preface to the Second Edition

When I was asked to revise *Setting Out*, the first edition had become rather dated and disjointed as additional information was added on the end in a later reissue. I have therefore taken the opportunity to reorganise the book into a more logical order which has included incorporating sections from Part Three of the 1981 reissue into the main body of the first part of the book. I have dropped much information which I now regard as obsolete — vernier theodolites, logarithms, etc. — and deleted references to imperial units. Also, with modern computing facilities, e.g. pocket calculators, desk top computers, it is no longer necessary to make many of the approximations which were common in the past.

Modifications to existing material include a step by step description of the reduction of level observations by the rise and fall method as well as height of collimation, additional details on precise taping techniques, different methods for reducing horizontal and zenith angle observations and new sample forms for the adjustment of traverses and braced quadrilaterals. Many of these changes reflect techniques which have long been used by land surveyors but which have not been commonly used by civil engineers on site. With the tighter specifications now called for in much setting out, especially with larger structures being prefabricated than in the past, I felt that these techniques, with their more rigorous field checking procedures, should be used by the modern engineer.

New material concentrates on the advances in electronics and computers now used on site and in the design office. Sections cover electronic theodolites and tacheometers and automatic recording and processing of data. A new section provides a guide to errors in electromagnetic distance measuring (EDM) equipment and some simple calibration techniques to supplement the existing details on testing levels and theodolites. The concept of digital terrain models (DTMs) is briefly covered and how they influence the derivation of setting out data.

I have tried to emphasise throughout these sections on electronics how the engineer must guard against adopting a 'black box' syndrome. Simply because this modern equipment is so easy to use the engineer must not forget the basics of good surveying. No matter how sophisticated and precise the equipment is, if the basic survey is badly planned, e.g. very narrow angles of intersection, a bad solution will result. With much of the burden removed from the actual measurement more time can be spent in planning what to actually measure. Similarly just because a field computer displays coordinates to a tenth of a millimetre the engineer must remember, and make allowances for, how good the original data were.

The book is still intended as a reference guide for the site engineer. It should also prove useful to diploma and undergraduate students in civil engineering, building or construction. Additionally it is recommended for undergraduate or recently graduated land surveyors. Basic site surveying and setting out are two subjects often inadequately covered in land surveying courses and yet many graduate land surveyors find themselves doing this type of work in their early employment.

D.M. Stirling
February 1988

Preface to the First Edition

The theoretical aspects of setting out new construction are not difficult to understand, but the practice requires a degree of skill and knowledge of practical methods which are not so well documented as some other aspects of surveying. This book is an attempt to provide knowledge of practical methods which will enable engineers to tackle setting out problems with confidence.

Since the process is really surveying in reverse, some notes on ordinary surveying practice are included. They are not a substitute for standard works, or training in the subject, but serve to remind the setting out engineer of the important aspects of the production of site plans so that any setting out plan made will start from a sound base.

The advice on instrument handling is a useful guide to the attainment of skill in use, without which the most sophisticated and expensive instruments will not produce good results.

S. G. B.
January 1975

1 Introduction

Setting out is of prime importance to the success of any construction work. It is vital, therefore, that the setting out engineer should have a clear idea of the task and should plan ahead for its proper execution. Often, too little attention is paid to the proper tools for the job and the order and method of carrying out the work. A good deal of planning goes into other aspects of construction and unless the setting out engineer clearly understands the task, plans it so that he or she is not continually overtaken by events, and executes it with the proper and timely provision of marks, information, profiles, levels, etc., he or she will not only lose the confidence of the remainder of the construction team, but may seriously prejudice the successful completion of the contract.

The following points are intended to go some way towards providing a clear view of the task and to assist young engineers in its planning and execution.

Instruments

A good engineer should be skilled in their use, should treat them with care, check them frequently, and should be capable of adjusting them.

Stores

Setting out requires proper equipment, as does any other site operation. The necessary stores should be organised and ready for use. (See check list on p. 6.)

Liaison with Site Staff

The engineer should know the planned sequence of construction and should find out what special information others may need so that a

1

suitable plan to meet all requirements can be made. He or she should make sure that the meaning and importance of all marks and pegs are understood by those who are going to use them.

Checks

It is important for the setting out engineer to work *from the whole to the part*, constantly checking the correctness of his or her own work and that of others which is based on this information. Pegs can be moved, sight rails altered or used with the wrong size of traveller, etc. If checks are not done, expensive mistakes may occur.

Records

Level books, field books and a setting out record book must be kept, with clear notes and diagrams so that what the setting out engineer has done can be clearly understood by others. Untidy and anonymous figures in field books and level books and calculations without clear notes and explanations are the mark of a poor engineer.

The Task

Setting out can be broken down into consideration of the following:

- Plan (or absolute) position
- Relative position
- Size
- Shape
- Level (in relation to MSL)
- Verticality

Facts that should be known:

- Accuracy required
- Data available and required
- Instruments available
- Time available

The Plan

This should be simple and should include the earliest provision of permanent marks and control for:

(1) Level (establishment of Temporary Bench Marks, properly marked and constructed in safe and convenient areas).

(2) Position (establishment of base lines or grids, checking of layout against the ground, establishment of reference pegs).

Plan Position

The engineer is always faced with the problem of accuracy in this respect and must view the project as a whole to determine the limits to which he must work, and the data available on which to base his measurements.

On restricted sites in urban areas, works may have to be located with very considerable accuracy if they are to fit. If the setting out drawing gives no clue, then a careful check of the site against the overall dimensions of the project will provide a measure of the care which must be taken in absolute location.

Motorway centre lines are rigidly controlled as to position and may be related to the National Grid in co-ordinate values for key points to very close tolerances.

Care is also needed in new housing developments involving very large numbers of houses since, by the time the layout reaches the engineer, it will have been exposed to many sources of inaccuracy, e.g. the expansion or contraction of the original, copying inaccuracies, plotting errors, etc.

The dimensions and separation of the various buildings are seldom in doubt, since they are detailed in separate drawings and in the specification, but it may well be that with the various factors described above, the plan dimensions and separations do not agree with the detailed drawings. Unless the engineer is on guard and carefully checks the layout against dimensions gathered from other sources, it may well be that 99 houses are set out where 100 should be! The importance of working from the whole to the part cannot be over-emphasised in this respect.

Always base setting out on original survey marks if they exist; often co-ordinate values are available for these, so that their relative positions can be checked despite printing errors. Failing this, work from points of permanent detail which will have been fixed accurately in the original survey and cannot have moved. Check their relative positions on the ground against the plan positions. The scale of the layout or setting out drawing will often give a lead and assistance in deciding on and obtaining the required accuracy. If it is a large scale then obviously the original survey was done to finer limits and setting out accuracy must be correlated with this.

Use detailed dimensions, or calculated ones wherever possible, and avoid scaling from the plan unless all else fails, but be sure to check for distortion and allow for it.

Always establish permanent control marks outside the area of operations at the very earliest stage, or much detailed work will have to be repeated and good control lost as work goes on.

Relative Position

Accuracy in this respect is often decided for the engineer, since detailed dimensions are involved. In deciding on the limits and methods to be used the engineer will take into account the nature of the work. Care must however be taken in relation to building lines to ensure that there is no encroachment.

In industrial complexes with many inter-connecting underground and overhead services, tolerances will be tight. The use of a grid in such cases is essential, so that running errors will not accumulate and all setting out will be speeded up. Stanchion bases or precast column bases should, in general, be set out to the nearest 2 mm between columns.

Size and Shape

These need little explanation and the details are shown in the drawings, but the engineer must *always* check diagonals for squareness, however competent and confident he or she may be.

Difficulties arise when the construction is irregular or curved, and in such cases recourse to detailed calculation may be necessary to ensure that the setting out marks faithfully interpret the contract documents and are sufficiently numerous and clear to be worked from by operatives. Offsets from an axis or axes, or deflection angles, etc., may have to be used.

Level

Accuracy in respect of level varies little and is more influenced by the method and the instruments than by the project. It is, generally, no more difficult or time-consuming to read a level staff to 0.002 of a metre at all reasonable distances than it is to read to 0.10. There is no excuse ever for being careless in the levelling of an instrument or the reading of a staff.

All setting to level must be checked back to source. The early establishment of properly checked TBMs in convenient parts of the site will make level control, and checks thereafter, an easy matter.

Ordnance Bench Mark values on maps should not be accepted, and should be checked against lists either in the local authority offices or direct from the Director General's Office of the Ordnance Survey, where the up-to-date revised values are held. This is particularly important in areas liable to subsidence. Wherever possible, site TBMs should be related by check levelling to more than one OBM.

A water level is a convenient device for carrying datums round large floor areas.

Transfer of levels from the ground to upper storeys should be done, using a steel tape, either outside or in stair wells or lift shafts, so that cumulative errors do not occur.

In setting pegs to level and in providing sight rails, it is good practice and of assistance to the construction workers if they are set to the nearest multiple of 100 mm above or below the required level. This is not difficult to do and helps to avoid the errors likely in measuring finicky amounts. This does not, of course, apply to pegs which have to be set at a definite level to control finished surfaces, screeds, etc.

Always apply the level test and adjust any errors; although allowance can be made for known inaccuracies, it is better to do the adjustment as soon as opportunity offers.

Verticality

On ordinary buildings and low structures this can be most easily controlled by plumb lines. The bobs should be heavy, the ordinary instrument variety is too light. On high rise structures and multi-storey work an instrument check is necessary. If no Autoplumb or laser is available, a check in two planes at right angles to each other is necessary. These checks are made in the plane of the faces of the structure and reference pegs in these planes must be established (and protected) at an early stage.

On tall buildings there is no substitute for an Autoplumb or laser if a disproportionate amount of time is not to be spent checking verticality. Consideration of the position of the grid must be made so that suitable apertures can be left in floor slabs for upward sights. Permanent marks on small plates let into the ground floor slab should be made as soon as it is constructed.

If an ordinary theodolite is being used for observations at high angles of elevation for the checks, it is most important that it be tested for dislevelment of the trunnion axis, and any error corrected. More than usual care should be taken when levelling the horizontal plate. The Autoplumb and some lasers have a compensating device for upward sights, but the tests described in the handling notes should be done, and any necessary corrections made.

Checks on verticality are often best done when construction work is finished for the day, since this eases the problem of communication between the engineer and the assistant, who may be ten floors up. A field telephone or walkie-talkie is a great asset in these conditions.

Setting Out Equipment Check List

Survey Equipment

Level
Level staves
Theodolite
EDM and correct prisms, if necessary
Steel tapes: 4, 30 and 100 metre
Tape repair kit
Tape grips and tension handles
Taping arrows
Ranging poles
Ranging pole tripods
Plumb bobs
Field and level books
Set of scales
Basic drawing instruments
Set of set squares
Circular protractor
Ink markers and marking crayons
Pocket knife
Nylon line (hanks)
Straight edge (steel)
Calculator

Tools and Materials

Claw hammer
Lump hammer
Sledge hammer
Jumper bar
Cold chisel
Rasp or 'Surform' plane
Rip saw
Boat level
Pegs (50 × 50) as required
Stakes (50 × 50 × 2 metre) as required
Timber (100 × 25) as required for sight rails, etc.
Nails (various)
Builder's square
Paint (primary colours, emulsion or acrylic)

Paint brushes
Oil and wiping rags

Miscellaneous

Stationery (as required)
Graph paper

Electronic Equipment

When using electronic equipment on site (calculator, EDM, electronic fieldbook, laser, etc.) the engineer should always check that the necessary cables are available and that the battery and spares, if necessary, are fully charged. Nothing is more annoying than having the battery run out at a critical moment. If nickel cadmium batteries are being used then special care is required when charging them. These batteries need to be cycled at regular intervals, that is, completely drained, fully charged and completely drained again. If they are continually topped up without cycling for a long period they loose their ability to hold a full charge.

The Setting Out Record Book

There is no formal layout for the setting out record book. The important thing is that it should exist and should be a concise and logical record of the setting out on a contract, which can be readily understood and used by anyone connected with the task. The following suggestions will be of assistance in compiling such a record.

Form

A large (A4 or foolscap size) stiff-covered book is useful for medium-sized contracts not involving a large amount of scheduling and calculation work. For large contracts a loose-leaf type file is better, since copies of calculations and schedules done by photocopying or computer print-outs can then easily be included.

Contents

Index Always make provision for this and keep up to date against numbered inserts or pages as the work progresses.

Basic Data   Devote the first section to a record of this, with information on such items as OBMs used, TBMs established, survey marks used, base lines, grids, etc., established with details of method, marking, location and so on.

Site Clearance   Record details of original marks, and levels taken with reference to numbered field and level books.

Roads   Curve data, IPs, TPs, calculations, vertical curve details, centre line and margin levels, super-elevation calculations, schedules, details of profiles, padstakes, reference pegs, etc.

Drainage   Details of MH pegs, reference offset pegs, profiles, traveller lengths.

Foundations   Marks, levels, profiles, reference pegs, etc.

Buildings   Corner pegs (offset details), floor levels, grid markings, vertical control, etc.

In the recording of the details suggested, common sense must be the guide, but the engineer should always try to set down the information so that it can be understood by colleagues or a relief without the need for further explanation.

Throughout, reference should be made to numbered field and level books where there are extensive data to record. When full, these books should be kept together with the setting out record as part of the contract documents. They may provide very valuable evidence in the case of extra payments, disputes, etc.

Field and Level Books

Both field and level books should be used constantly to plan and record what is done. It is essential to cultivate method, neatness and good habits in this aspect of site engineering right from the start. The ideal to be aimed at is that any fellow engineer should be able to consult your field and level books at any time and see:

• Who did some operation or measurement
• Where it was done
• When it was done
• What was done
• How it was done.

From this information it should be possible to carry the work to a further stage without the necessity for remeasurement or checks of accuracy.

To assist in the proper keeping of these books two or three pages should always be left at the front for an index of the contents. Inside the cover, the following information should be recorded:

* Your name and status
* The date taken into use
* The serial number (your fd book number)
* The name and section of the contract.

Pages should be numbered and the index kept up to date.

Diagrams should be used liberally both to keep your own mind and recollection clear, and to assist in understanding by others. One of the most informative and simple ways of conveying this kind of technical information is just to say what you did, e.g.

Set up at original survey peg 7, laid on peg 8, turned off 14°40′20″ for line of main road tangent. Distance peg 7 to chainage 17 + 50. 31.56 m.

Do not be afraid of appearing over-methodical or pedantic. An un-methodical engineer is a menace on any contract. Remember that two minutes after anonymous figures have been recorded no one else will know what they mean. Two days later you will not know either.

Reminders

Know the layout.

Get out on the ground and become thoroughly familiar with the site.

Work from the whole to the part.

Work to the accuracy required.

Establish permanent marks.

Keep clear records.

Look and plan ahead.

Always check your work.

Part One
The Survey

2 Instrumentation and Observing Procedures

Optical Systems

Many surveying instruments, e.g. levels, theodolites, sextants and optical plummets, use some form of telescope. The main viewing telescope consists of two distinct optical systems. These are the eyepiece system and the objective system.

The eyepiece system is focused by turning the knurled ring of the eyepiece cover to bring the image of the cross hairs of the reticule into sharp focus. Once this has been done for one's individual eyesight it should not need to be readjusted during operation.

The objective system is focused until the image of the object viewed (target, staff, etc.) appears in sharp focus in the same plane as the image of the crosshairs. Then a reading can be taken.

If either focusing is not done correctly, slight movements of the eye relative to the eyepiece will produce an apparent movement (parallax) of the cross hairs against the target or staff. If the eyepiece has been focused correctly, parallax can be removed by slight re-adjustment of the main objective focusing. If this is not successful the eyepiece should be re-focused separately against a neutral background (the sky or booking sheet, *not* the image of the target or staff). Focusing should always be done in this order. It is useful when an instrument is in regular use to note the observer's eyepiece setting (marked in diopters on the eyepiece ring) so that correct focus can be set rapidly.

The most economical combination of lenses in the objective system results in an inverted image of the staff or target. However observers soon become used to this inverted image. Modern instruments may have an additional lens incorporated in the objective system to produce an erect image.

Many levels and theodolites have an additional secondary telescope for reading the glass circles or scales incorporated in these instruments. In these telescopes only the eyepieces can be focused onto the plane of the index mark. Again the focusing should be checked for parallax which can produce errors in the scale reading.

13

Reticules

In all modern level and theodolite main telescopes the reticule is engraved on a glass diaphragm. All reticules are variations of a simple cross (Fig. 2.1). The main horizontal hair spans the field of view and is the reading mark for the vertical circle or staff reading. The vertical hair is the reading mark for the horizontal circle. There are usually two shorter horizontal hairs above and below the main horizontal hair. These are called the stadia hairs and are used for distance measurement by stadia tacheometry. It would be very rare to find an instrument today where the stadia multiplying constant was not 100. It is sometimes possible to make a mistake in reading a staff by confusing a stadia hair with the centre hair.

Levelling

The levelling process is a simple one which makes use of a combination of a telescope fitted with cross hairs to denote the optical axis and a spirit level or compensator to enable it to be levelled. A graduated staff viewed through the telescope measures the height of the line of sight above the

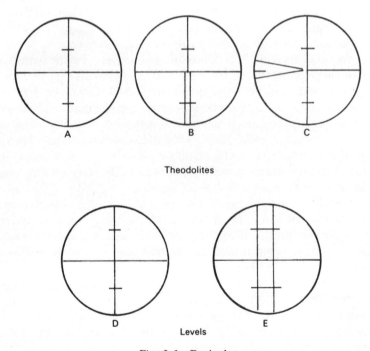

Fig. 2.1 Reticules.

foot of the staff. With the line of sight continuously level, differences in readings of the staff obviously correspond with differences in level of the staff positions. Levels can be arbitrary values for local work, but in general all works are related to Mean Sea Level.

Levelling Instruments

There are three main types of level with which the site engineer may come into contact:

- Dumpy level (Fig. 2.2)
- Tilting level (Fig. 2.3)
- Automatic level (Figs 2.5 and 2.6).

Other names are often used, e.g. Engineer's level, Quickset level, etc., but these descriptions are not accurate and the levels they refer to will be found to be one of the three main types. Quickset levels are so called because they have a ball and socket attachment to the stand.

Fig. 2.2   Watts microptic level (dumpy type) with horizontal optical circle.

Dumpy Levels

These have the telescope and levelling plate cast in one piece. They can be levelled only by use of the footscrews. The base containing the footscrews is known as the tribrach, as is the base of the theodolite. A large bubble is mounted parallel to the telescope (it has adjusting capstan screws). Often a smaller cross bubble is mounted transversely to assist in coarse levelling.

Once level, the line of sight is level in all directions. This is an advantage when a large number of intermediate sights has to be taken.

Tilting Levels

These have the telescope mounted on the levelling plate so that it can be tilted fore and aft by turning a screw which works against a spring-loaded plunger. Sometimes this screw takes the form of a gradient drum, so that the line of sight can be tilted to a given gradient. This can only be used if the level is first levelled at zero tilt by adjusting the footscrews or the quickset mounting, whichever is applicable.

A small circular (pond) bubble is usually mounted on the base to assist in coarse levelling. In use *the main bubble must be levelled for every reading*, and it is important to develop the correct routine in handling to avoid forgetting to do this. The main bubble may have two different arrangements for viewing from the eyepiece end:

• An angled mirror
• A split bubble display.

Fig. 2.3 Watts tilting level with quickset mounting.

With the angled mirror a small flap over the bubble hinges up and clicks into a 45° position so that the mirror on its underside provides an image of the bubble visible from the observing position.

With a split bubble display the bubble is encased in a box with a small lens eyepiece. The view presented in this eyepiece is in the form of one end of a normal bubble with a line down the middle. It is a composite image of the two ends of the bubble and in use the instrument is level when the two halves present the image described. When the instrument is tilted, the image splits into two halves which run in opposite directions. If the coarse levelling is too coarse there may be no image visible at all, and a good deal of twiddling of the levelling drum will then have to be done to make it run into the field of view (see Fig. 2.4).

Automatic Levels

These may not resemble an ordinary level in appearance at all, being more like a box or a camera in appearance. They have a levelling base with a circular bubble or two opposing small bubbles for coarse levelling, which is all it is necessary to do before use.

Inside the telescope the line of sight passes through a prism system with a counter-weighted compensator. Any departure from a truly level line of sight after coarse levelling is done is taken up by the compensating prism which refracts the line of sight so that it emerges from the object glass horizontally. In practice, automatic levels are quick to use in 'no wind' and vibration-free conditions. Wind or nearby plant working often causes shiver of the cross hairs which makes staff reading more difficult.

The system of compensating only works within limits and the coarse levelling must not be too coarse. A slight tap on the instrument after levelling will cause the hair lines to float, showing that the compensator is working. The light path through the system is such that automatic levels have erect, not inverted, images.

Levels come in a variety of cases designed for safe transport. When lids are opened or covers removed always take note of the position of the instrument and the holdfast arrangements so that they can be duplicated when putting it away.

Level

Un-level

Fig. 2.4 Split bubble display.

Fig. 2.5   Watts automatic level.

Level Stands

There are three main types of level stand:

(1) Solid leg, non-telescopic
(2) Telescopic
(3) Braced framework legs, non-telescopic.

The telescopic is probably the most common, though nearly all old dumpy instruments have solid leg stands. Telescopic level stands are less stable than the solid or braced framework leg types, but are easier to transport.

There are two main ways of fixing the level to the stand; in both cases the underside of the levelling plate has a threaded aperture. On old dumpy levels, the top of the stand has a large screw thread protected by a cap, which is removed so that the level can be screwed on. In more recent instruments there is a captive bolt projecting through the top of the stand, which can be rotated to screw into the levelling plate. There is some degree of standardisation of thread size, but stands should always be checked against instruments before use. Some level and theodolite stands are interchangeable. Telescopic stands have sliding portions which can be clamped in any convenient position. Undue force should not be used when clamping; finger-tightening is enough.

Leg hinges and the bolts fixing the wood portions to the metal should be inspected for tightness. Metal shoes on the legs should be similarly inspected. Undue looseness in any of these fittings will cause difficulty in

Fig. 2.6 Kern automatic level.

levelling the instrument and may produce errors due to the instrument going out of level in use.

Setting Up

A systematic method will develop speed and accuracy.

Stand

(1) Undo keeper strap.
(2) Unclamp legs.

(3) Extend legs evenly and re-clamp.
(4) Grasp two legs near the head and open out into a convenient position.
(5) Plant legs keeping head level. On soft ground press feet lightly into soil. On hard ground ensure that points are not on a sloping surface.
(6) Stand back and view head for level; adjust if required by moving one leg.
(7) Unscrew cap or cover plate if fitted.

Level

(1) Open box (note position of level and clamping arrangements).
(2) Undo clamps and remove level by grasping telescope tube firmly.
(3) Screw on to stand (never leave go of the instrument until firmly fixed).
(4) Remove telescope cap, fit or pull out ray shade, put cap in box.
(5) Close box, put in safe place.
(6) Undo slow motion clamp.
(7) Set tilt drum if fitted, to zero.
(8) Focus eyepiece.
(9) Level instrument as described below.

Levelling the Instrument

Except on quickset levels, the levelling screws on the base or tribrach are fitted with large knurled rings so that they can be turned easily. They terminate in ball and socket joints in the lower part of the plate and work in threaded portions of the upper part. They form an equilateral triangle.

The level is turned so that the telescope is parallel with the line of two of the footscrews. These two screws are then grasped lightly between finger and thumb of both hands, and turned *equally* and *simultaneously* in *opposite* directions. This tilts the plate, causing the bubble to run; it always follows the direction of movement of the *left thumb*. With this knowledge it is possible to turn the screws the correct way initially by noting the position of the bubble. Levelling is a delicate operation and the hands and the remainder of the body should be kept clear of the instrument and stand.

When the bubble is centred by final deft touches of the two screws, the telescope is turned through 90° so that its axis lies over the remaining footscrew. It is now centred in this position by movement of the remaining footscrew.

If this whole operation has been done correctly and there is no mal-adjustment of the bubble, it will remain central wherever the telescope is turned. In practice this seldom occurs and a final stage of levelling is required. This is done as follows.

The position of one end of the bubble is noted against the graduations of the bubble tube, and it is then turned end for end, and the *same* end read again. If there is a difference the bubble is relevelled to make the same end read the mean reading, and the telescope again turned end for end to check that the mean position is maintained. Now the telescope is turned through 90° so that it is again parallel with the line of two footscrews, and the bubble set to the mean reading by *simultaneous and equal movement* of the two screws. It should now take up this mean position wherever the telescope is turned. This is known as 'meaning the bubble' and produces a level line of sight in the telescope. It can only be done with dumpy levels.

For tilting levels it is sufficient to carry out the first steps to centre the circular bubble. All subsequent shots are preceded by levelling the main bubble with the gradient drum once the staff has been focused.

It is important to develop a systematic method of aligning, main telescope focusing, levelling, reading and checking to guard against reading and other errors. The sequence should be as follows:

(1) Rough alignment, clamp
(2) Final alignment, with slow motion screw
(3) Main focus, check for parallax
(4) Check bubble
(5) Level bubble (tilting levels)
(6) Read staff
(7) Book reading
(8) Check bubble
(9) Check reading
(10) Check booking
(11) Signal staff holder to move on.

In steps (6) and (9) it is important to check the whole metre figure in the field of view.

Reading

The normal inverted image presents initial difficulties in reading the staff, but these are soon overcome with experience. It is best to develop systematic habits to minimise the making of mistakes.

The following sequence is recommended:

(1) Identify the whole metre and tenths of metre
(2) Read the ten millimetre division
(3) Estimate the individual millimetres, if required
(4) Book the reading
(5) Check the reading
(6) Check the booking.

Staff holders should be trained to hold the staff with the palms of the hands at the side of the staff so as not to obscure the graduations with their fingers. Wherever possible staves with handles should be used as they are easier to hold steady.

Levelling Staves

There is a wide variety of staves on the market in terms of face markings. The most common version is shown in Fig. 2.7.

Staves are made of wood or light alloy and can be telescopic, folding or jointed.

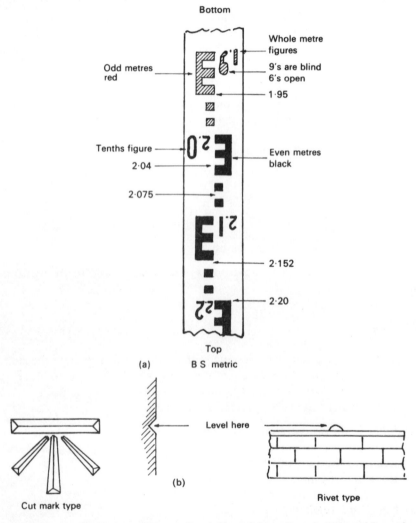

Fig. 2.7 (a) Typical levelling staff markings (viewed as an inverted image).
(b) Typical Ordnance Survey benchmarks.

Markings

The British Standard markings are sometimes known as the 'E pattern' as the marks in every tenth of a metre resemble the letter E. The staff is divided into metres, tenths and hundredths. Marks are 10 mm wide and the spaces between are the same. The *bottom* of the tenths graduations mark the whole metres and tenths.

The lowest direct reading is 10 mm. It is possible to estimate finer readings, at first to 0.005 and with practice to 0.002 and 0.001 of a metre. The marks are in red on the odd metre lengths and black on the even metres. The graduations are figured at every tenth of a metre and the whole metre figures are noticeably smaller than the tenths, e.g. 0.8.

Booking

There are two methods for booking and reducing staff readings, the rise and fall method and the height of collimation method.

The Rise and Fall Method

Figure 2.8 shows a typical fieldsheet layout for this method. It is important that all the header information is filled in so that the results can be properly interpreted. Additionally the right-hand side of the page allows each horizontal line to have notes relating to the entry or reduced level. Intelligent use of this page, without slavish following of the principle of one line, one piece of information, may make all the difference to the understanding of the results by another person.

Sequence of Completion

The sequence of completion, using the level survey from Fig. 2.9 as an example, is as follows:

(1) Fill in the header information. If the observer is booking his or her own observations then the observer's name or initials should be entered on both lines.
(2) Enter reduced level of starting level.
(3) Note description of starting level in 'Remarks' column.
(4) Enter first backsight on *same horizontal line*.
(5) Enter any intermediate sights on *successive horizontal lines*. Note against each line the significance of the sight in the 'Remarks' column.
(6) While the staff holder is moving between positions reduce readings. Subtract each staff reading from the backsight or intermediate sight on the line *immediately* above it. If the result is positive then this is a

LEVELLING OBSERVATIONS

DATE ___4/3/88_____   LEVELS FOR _Checking O.S.B.M.s_
OBSERVER ____R.P.T_____   INST. TYPE AND NO. _NK10 224923_
BOOKER _____R.P.T_____   WEATHER __Sunny, some shimmer_

BACKSIGHT	INTERMEDIATE SIGHT	FORESIGHT	RISE	FALL	REDUCED LEVEL	REMARKS
1.791					26.94	O.S.B.M. by gate
	2.163			0.372	26.568	A
	1.970		0.193		26.761	B
	1.173		0.797		27.558	C
	-1.605		2.778		30.336	D (Invert on bridge)
1.833		0.911		2.516	27.820	E (C.P.)
	25 1.652		0.181		28.001	F (Transposed No.)
	1.568		0.084		28.085	G
		1.009	0.559		28.644	O.S.B.M. on church should be 28.65
3.624		1.920	4.592	2.888	28.644	∴ misclosure = -6mm
-1.920			+2.888		-26.940	∴ OK
1.704 ✓			1.704 ✓		1.704 ✓	

Fig. 2.8   Rise and Fall reduction of level observations.

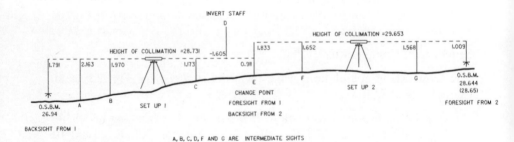

Fig. 2.9   A typical level survey.

'Rise'. If the result is negative then it is a 'Fall'. Enter this value in the relevant column on the *same* line as the reading. Add a rise or subtract a fall from the previous reduced level and enter the result on the same line as the reading. Note that the intermediate sight to point D is an 'invert' level, i.e. the staff was held upside down on the underneath of an object such as a bridge. Enter this type of reading with a negative sign, with the relevant remark, and reduce as above.

(7) Enter foresight and reduce as above.

(8) Move instrument to new position and observe and enter backsight *on the same horizontal line as the previous foresight*. Note that this is a change point (CP) in the 'Remarks' column.

(9) Repeat steps 5 to 8 until the line is closed on a suitable bench-mark. NEVER leave a level line unclosed. If no other suitable benchmark is available then the line should loop back to close onto the opening benchmark, otherwise gross errors in staff readings will go undetected.

At the bottom of *each* level sheet the following checks must be carried out:

(10) Sum all the backsights, foresights, rises and falls on the sheet.

(11) Check that the arithmetic on the sheet is correct by subtracting the sum of the foresights from the sum of the backsights, the sum of the falls from the sum of the rises and the first reduced level from the last reduced level. If the three answers are different then there is an arithmetic error on the sheet. It is important that all the subtractions are done the correct way as the signs of the differences are important especially if it is a closed loop level circuit.

(12) Check the final reduced level for accuracy against the known value of the benchmark. Enter the result in the 'Remarks' column. If this discrepancy is too large and all the checks in step 11 have agreed then the line will have to be reobserved.

It cannot be too strongly emphasised that neat small figures are required in booking owing to the small space available and the need for legibility and accuracy.

Mistakes in reading or reduction are sometimes made. It should be the golden rule that mistakes in reading are rectified by crossing out the incorrect figures and writing in correct ones, *never by alteration or erasure*. Mistakes in reduced levels or entries other than staff readings can be corrected by erasure and substitution.

To summarise the rules for good booking:

(1) Information must be written in at the start.

(2) A separate line must be used for each level.

(3) A label for each level.

(4) Corrections to readings, *not* erasures.

(5) BS/FS check, rise/fall check, last RL/first RL check.

(6) Neat, small and legible figures.

Height of Collimation Method

Figure 2.10 shows a typical height of collimation method fieldsheet with
the results from the same survey as above.

The recommended procedure is:

(1) Fill in the header information.
(2) Enter reduced level of starting level with a description in the
    'Remarks' column.
(3) Enter first backsight on the *same horizontal line.*
(4) Add this reading to the starting reduced level and enter the result in
    the 'Height of Collimation' column on the *same horizontal line.* This
    gives the level of the horizontal plane defined by the instrument.
(5) Enter any intermediate sights on *successive horizontal lines.* Note
    against each line the significance of the sight in the 'Remarks'
    column.
(6) While the staff holder is moving between positions reduce reading.
    Subtract each reading from the 'Height of Collimation' value from

LEVELLING OBSERVATIONS

DATE ____4/3/88____   LEVELS FOR _Checking O.S.B.M.s_
OBSERVER ____R.P.T____   INST. TYPE AND NO. _NK10 224923_
BOOKER ____R.P.T____   WEATHER _Sunny, Some shimmer_

BACKSIGHT	INTERMEDIATE SIGHT	FORESIGHT	HEIGHT OF COLLIMATION	REDUCED LEVEL	REMARKS
1.791			28.731	26.94	O.S.B.M. by gate
	2.163			26.568	A
	1.970			26.761	B
	1.173			27.558	C
	-1.605			30.336	D (Invert on bridge)
1.833		0.911	29.653	27.820	E (C.P.)
	1.652 ²⁵			28.001	F (Transposed figs.)
	1.568			28.085	G
		1.009		28.644	O.S.B.M. on church
					(should be 28.65)
3.624		1.920		28.644	∴ misclosure=-6mm
-1.920				-26.940	∴ O.K.
1.704 ✓				1.704 ✓	

Fig. 2.10 Height of Collimation reduction of level observations.

step 5 and enter results in the 'Reduced Level' column. Note how the invert level to D is treated as a negative value as in the rise and fall method above.

(7) Enter foresight and reduce as above.
(8) Move instrument to new position and observe and enter backsight *on the same horizontal line as the previous foresight*. Note that this is a change point (CP) in the 'Remarks' column.
(9) Repeat steps (5) to (8) until the line is closed on a suitable benchmark.

Again arithmetic checks should be carried out at the foot of each level sheet. These are:

(10) Sum all the backsights and foresights on the sheet.
(11) Check that the arithmetic on the sheet is correct by subtracting the sum of the foresights from the sum of the backsights and the first reduced level from the last reduced level. If the two answers are different then there is an arithmetic error on the sheet.
(12) Check the final reduced level for accuracy against the known value of the benchmark. Enter the result in the 'Remarks' column. If this discrepancy is too large and all the checks in step 11 have agreed then the line will have to be reobserved.

Comparison of the Two Methods

The height of collimation method is the one which is more widely used on civil engineering sites. As a result it is often employed when it is not really suitable. Its popularity is probably due to the fact that, if there are many intermediate sights, the arithmetic is easier than for rise and fall. However the reduced levels of the intermediate sights *are not* included in the arithmetic checks. Reduced levels for intermediate sights *are* checked arithmetically in the rise and fall method. Therefore if it is important that all intermediate sights need to be known with good reliability then rise and fall *must* be used.

Unless the book is used only by the person booking, any level run should be signed by the person doing it. Examples of various types of booking for different purposes are shown in the diagrams in this section, and in some others, e.g. under level gridding. On contracts where there are a number of engineers working, loose leaf level books are very useful for records, copies, etc.

Levelling Practice

The following instructions concern important parts of practice which must be learnt and used if skill is to be developed. A modern, properly adjusted instrument still needs skill in use to produce correct results.

In check levelling always keep backsight and foresight distances equal to quite fine limits. Keep this constantly in mind when choosing new positions for setting up. Do not allow the staff holders to go where they like and train them to keep to these limits by rough pacing to and from instrument when changing staff position to a change point.

Restrict sighting distances to about 25 m. Despite good telescopes it becomes difficult to make accurate staff readings much beyond this distance. For intermediate sights for spot levels where required accuracy may be less, this rule may be relaxed.

The staff will only show a true vertical distance when truly vertical. Unless it is fitted with a bubble, it is not possible to know when this is so. To overcome this the staff holder should be trained to rock the staff slowly back and forth, *when signalled to do so*. The staff needs to be still when being focused. The rocking causes the hair line of the reticule to make an apparent movement up and down, the lowest reading showing when it is truly vertical. It is not difficult to read this low reading and thus be sure that error has not been introduced.

The level staff should always be placed on a firm foundation at change points, if possible a point support. This will ensure that there is no change in level when it is reversed for a fresh backsight reading. When rocking, this point support is also important. A large based staff will rock up on one edge on a flat unlevel surface, giving a false low reading.

Do not allow the staff holder to lay the staff down at a change point while waiting for you to move and set up again.

On soft ground a large stone pressed in by the heel will be a better change point than the ground. There are special crow's-foot devices with a chain attachment for the staff. Their use is worth considering when establishing a number of temporary bench marks over a large area of undeveloped ground.

Check the verticality of the staff in the visual plane and signal the staff holder to correct if not right. Always check back to source or between benchmarks of known validity. Always ensure that any point required for future reference (e.g. a TBM) is a change point so that it is included in the automatic BS/FS check.

Always treat the level gently and avoid contact with it other than by the tips of the fingers of the operating hand. On soft ground avoid undue movement around the legs of the instrument.

With tilting levels always make the bubble level *immediately before reading* the staff, and check afterwards.

With a dumpy level, the bubble may run out for some reason as a level is about to be taken; a touch on the appropriate footscrew will correct this without seriously altering the height of instrument.

The reduction of levels during the run of observation is a useful check against error and should be practised. A routine must be developed to do this only when the staff holder is moving, to avoid delays.

Never accept that a level is in correct adjustment unless you have tested it. Learn how to carry out the level test, and always apply it to a level the first time used (Fig. 2.11).

By setting up in the centre of a base AB the absolute difference in the heights of A and B is obtained, any dislevelment in the line of sight one way being cancelled by an equal amount in the other.

Observations from outside the base at C reveal the nature and extent of any error which can be then corrected to horizontal.

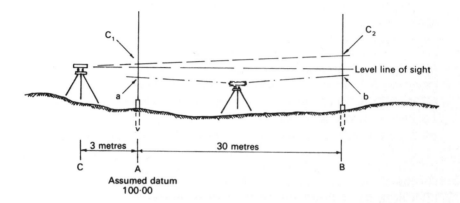

Booking example

BACKSIGHT	INTERMEDIATE SIGHT	FORESIGHT	RISE	FALL	REDUCED LEVEL	REMARKS	
(a) 2·60					100·00	A	) FROM MID
		(b) 2·79		0·19	99·81	B	) POINT
(c₁) 2·82					100·00	A	) FROM
		(c₂) 3·13		0·31	99·69	B	) C.
2·81					100·00	A	) CHECK AFTER
		3·00		0·19	99·81	B	) ADJUSTMENT

Level of B from C is 99·69, i.e. 0·12 lower than the absolute difference, showing that the line of sight from C is canted upwards, giving too high a reading (3·13) at C_2.

To correct, the telescope is canted down to give the correct reading. This is done by taking 11/10 of the difference = 0·13 (CA is to AB as 1 to 10).

This will lower reading C_1 by 1/10 = (0·01), giving new readings C_1 and C_2 of 2·81 and 3·00, giving correct height for B of 99·81.

Fig. 2.11 Levelling − the 'Two Peg' test.

On long runs, always record the levels of certain easily identified points, so that if errors occur the effect can be localised and corrections made without the necessity for repeating the whole run.

Never move the instrument unless the staff holder is on a change point.

General Care of Instruments

(1) Always carry in box during vehicle transit.
(2) Always keep legs folded and strapped in transit.
(3) Replace instrument in box with clamps secured when not in use.
(4) Always centralise footscrews before replacement.
(5) Dry before storing after exposure to rain.
(6) Wipe lens *only* with a tissue or dusting brush.
(7) Centralise slow motion screw from time to time.
(8) Always stand up vertically with legs spread when in use, never lean against a wall or vehicle.
(9) Never leave unattended when set up, particularly where there is site traffic.
(10) Check tightness of leg hinges and metal fixings.
(11) Keep cover plate or cap of stand fixing secure when not in use.
(12) Check that adjusting tools are secured in box.
(13) Most levels are supplied with a container of silica-gel crystals to absorb moisture in box. These crystals turn pink when they have reached saturation. Dry out on a radiator or space heater to drive off moisture and replace in box. This prevents the growth of fungus on the optical system, particularly in damp warm climates.

The Theodolite

Types of Theodolite

Most modern theodolites are of the optical plate type with graduations marked on a glass circle or micrometer inside the body. They are usually read optomechanically through small auxiliary microscopes alongside the main sighting telescope. Some are of the direct reading type where the lowest sub-division of the scale is read by estimation to the nearest 1 or 2 minutes of arc, but most are of the micrometer type where it is necessary to set an index by turning a micrometer head before the reading of the odd minutes and seconds can be done. In recent years optoelectronic systems have been developed to automatically scan the glass circles. The result of the scanning is then displayed electronically on a LED or LCD display. These instruments are generally known as *electronic theodolites*.

(a)   (b)

Fig. 2.12 (a) Hilger and Watts ST200 optical circle micrometer theodolite.
(b) Wild T2 mean reading optical circle micrometer theodolite.

Apart from the various makers' model numbers it is usual to describe a theodolite by the lowest direct reading possible, e.g. a 20″ instrument would mean one capable of direct reading to 20 seconds of arc. Such an instrument is adequate for most setting out work, though on motorway work or where long sights are necessary a 1″ model would make some work more easy.

Nearly all makes share the familiar three-screw levelling tribrach arrangement, upper and lower plate clamp screws and slow motion tangent screws. Some models have no bottom plate clamp screw and tangent screw and the bottom plate remains steady by friction unless the screw locking both plates together is in the clamped position. These models are somewhat more difficult to use for setting out owing to this arrangement, and cannot be used for what is known as the reiteration method of observing small angles very accurately.

All types measure horizontal angles clockwise from 0° to 360° and it is useful to have a mental picture of the plate graduation with the 0° graduation opposite to the observer's position and facing the same way as the telescope. At least one optomechanical model has a facility for reading from 0 to 360 anti-clockwise by manipulation of a changeover knob. Nearly all electronic theodolites have this facility.

Electronic theodolites also have the facility to change the angular units which are being used in the display of the circle reading. There are three commonly used systems:

(1) degrees minutes and seconds,
(2) degrees and decimals of a degree, and
(3) gons.

It is very important that the units being displayed are the units which the observer is expecting to be displayed. This is a very simple mistake to make but one which produces enormous errors in the results.

Vertical circles are engraved in a number of different ways, some in full circles and others in quadrants, and it is best to check this, should the necessity of observing vertical or zenith angles arise. On some optomechanical types the images of the two circles appear in the same field of view of the reading telescope with the distinguishing marks H or V or Az and V against the main scale. The same setting micrometer, where applicable, must be used for both circles in turn. On electronic theodolites the readings of both circles may be displayed simultaneously or it may be necessary to select each display in turn by means of a switch. Good engineers will take pains to become thoroughly familiar with their instruments. A maker's handbook can always be obtained if there is not already one with the instrument.

Care is needed to guard against accidental reading of the vertical scale, particularly with some instruments where the vertical circle is arranged to give values close to 0° or 90° for small angles of elevation at face left.

There is some degree of standardisation in the arrangements for fitting to the stands but when there are a number of instruments in use on a contract it is always best to check that the correct stand is kept with its own instrument.

Most instruments are fitted with an optical plumbing device consisting of a small telescope in the body of the instrument with the line of sight turned through 90° to look vertically downwards when plumbing over a mark. There is a small reticule showing cross hairs in this field of view which must be focused to a sharp image in use by manipulation of the eyepiece. It is a very useful device, particularly in windy conditions or when plumbing over marks which are either very near or deep down. Note the later remarks about instrument testing in this respect.

The following is intended as a guide to the development of good habits without which it is difficult to become really skilled in instrumental measurement.

Setting Up

Undo the leg keeper strap, extend the legs equally if telescopic and clamp firmly but not tightly.

Grasp two legs and plant the tripod roughly central above the ground mark, keeping its head level; stand back and view for level against the horizon; correct if required.

Drop a small stone and observe distance and direction of centring error; correct by moving two legs and pivoting about the third. When satisfactory, press shoes of two legs firmly into the ground, check level of head and correct if required by moving the point of the third leg in a small arc, not in or out (towards or away from the mark); press in shoe.

Remove the protecting cap from head of tripod and secure to purpose-made fitting or to the leg keeper strap. Open instrument box and note how the instrument is secured.

Undo clamps, grasp firmly by A frames (alidade) and mount on tripod, either by rotating the tribrach so that the instrument screws itself onto the thread provided, or, in other types, by rotating the fixing screw so that it screws into the bottom of the tribrach.

Never release the grip of the A frames until secured

On instruments fitted with a sliding attachment for centring fine adjustment, unlock the clamp and centre.

Attach the plumb bob and accurately centre the instrument by minor pressure on the shoes or by use of the sliding attachment; clamp when satisfactory.

Centring with Optical Plummets

The following method will overcome any difficulty experienced with centring by the use of the optical plummet now almost universally fitted to theodolites.

With the instrument set up approximately over the mark and the tribrach reasonably level, look through the optical plummet eyepiece (correctly focused) and, by manipulating the footscrews, bring the cross hairs or centre onto the mark without regard for level.

Turn the instrument so that the plate bubble lies parallel with the shoes of two of the legs. Centre the bubble roughly by pressing down on the shoes or by using the telescopic leg joints. Turn the instrument through 90° so that the bubble axis points to the third leg. Level in this plane by the same method. Fine level the bubble in the normal way by the use of the footscrews.

It will now be found that the optical plummet centre is very close to the mark with the instrument level and well within the minor adjustment of the sliding head.

Remove the lens cap from the instrument and place in its box, close the box and put in a safe place.

Levelling the Horizontal Plate

Before attempting to level the instrument all tripod wing nuts should be finger-tight, the ray shade fitted, eyepieces focused and the slow motion screws brought to the centre of their runs.

Turn the plate so that the main bubble is parallel to the line of two of the footscrews. Centre the bubble by simultaneous movement of the two screws in opposite directions. The bubble will move in the same direction as the left thumb.

Turn through 90° so that the bubble now lies parallel to the line of the third screw and the centre of the others; centre the bubble using third footscrew.

Turn through 180° and check that bubble stays central. If it does not, correct half the error and turn again through 180°. Bubble should now take up mean position.

Turn over original screws and bring bubble to this position by equal and opposite movements of the screws. Bubble should now remain still wherever instrument is pointing.

Note: The accuracy of levelling the plate should be related to the task in hand. For accurate work involving high angles of elevation or depression the whole process may have to be repeated, but for all normal horizontal measure in setting out levelling within one division is sufficient.

Centring over the Groundmark

Release the thread securing the theodolite to the tripod by about half a turn. (On theodolites with large thread fittings there should be a special clamp which can be released to enable the top portion of the instrument to translate with respect to the tripod.) Look through the optical plummet and then gently move the theodolite across the tripod head to centre the plummet reticule over the groundmark. When moving the theodolite be careful to move it along two mutually perpendicular axes, one parallel with the two footscrews which were used together for levelling and the other perpendicular to this axis. *Do not* rotate the theodolite on the tripod head as this will dislevel the instrument. One particular instrument manufacturer, Zeiss (Oberkochen), has a special system which uses two perpendicular grooves on the base of the tribrach which fit over two pins on the tripod head, inhibiting any rotation of the instrument on the tripod head. Tighten the thread fastening the theodolite to the tripod.

The result of this procedure will be that the theodolite will have gone slightly out of level. There should now follow an iterative procedure of relevelling and recentring until the instrument is levelled and centred satisfactorily.

Sighting

Check that the image of the cross hairs is sharp and clear, note diopter reading on eyepiece for reference in case of accidental movement later.

When laying on a mark, sight on the lowest visible portion. Align the telescope roughly by use of the open sights so that the mark will be visible when viewed through the telescope. Bring the image into sharp focus with the main focusing knob or ring.

Using the slow motion screw, accurately bisect the mark in a single smooth motion. With spring-loaded tangent screws it is advisable always to do this by a turning motion, that is compressing the spring. Remember to unload this compression before the next sighting. Do not make finicky back-and-forth movements on the point of bisection of the mark; if not satisfied with the first result, turn right off and bring up again. This method helps to eliminate any errors due to backlash in the slow motion gear. Check for parallax before and after sighting, refocus if required using main focus only.

The lightest of finger and thumb manipulation is required at all times if good results are to be obtained and care must be taken to leave the instrument undisturbed by contact with any part of the body or clothing. Movement round the instrument should be restricted to that which is necessary for use.

Measuring Angles

It is normal practice to measure angles on both face left and face right as this eliminates many small instrumental errors without the need to carry out the tedious instrument adjustments described later. It is also good practice to simply *read* the direction to a suitable reference object (RO) rather than to attempt to *set* a particular circle reading. On many electronic theodolites, however, it is a simple matter to enter a specific circle reading using the keyboard of the instrument. Many electronic theodolites also have the facility for measuring, and then subsequently correcting for, instrumental errors. In this instance a well-defined target is observed on face left and face right. The microprocessor in the theodolite calculates what errors produced the discrepancies between the two readings and then calculates corresponding corrections for all subsequent measurements which only need to be observed on a single face.

There are two methods which are appropriate to setting out:

• The repetition method (for small angles)
• The direction method.

In the repetition method the readings to the two marks between which the angle is required are taken, then, instead of changing face, the last reading is left on the plate, the bottom plate is unclamped and the instrument relaid on the RO using the bottom plate tangent screw. The second mark is then bisected again using the top plate: this has the effect of doubling the angle, which is not read this time. The process is repeated, say, twice more, and the final angle deduced by dividing the result by the number of times observed. Errors of reading are thus reduced; the method is of value for small angles.

In the direction method, reading is commenced by sighting the RO with the plate reading somewhere between 0° and 5°, and each mark to which an angle is required is bisected in turn and the readings noted. At the final mark the telescope is transitted, the plate turned through 180° and each mark again intersected. The readings will naturally differ by 180° on face right. The close on RO should be within acceptable limits and the mean value used for subtracting from the means of all the readings to the other points. For further rounds the RO reading should be arranged to fall on a different sector of the plate each time.

The method can be varied by closing on the RO on each face though this takes slightly longer to do. It is useful for a beginner as it detects error at an earlier stage of the operation.

Figure 2.13 shows examples of booking the readings.

The repetition method is not possible on most electronic theodolites.

Vertical and Zenith Angles

Although the term vertical angle is widely used to describe readings from the vertical circle it is probably better to distinguish between two specific types of angle − *zenith angles* and *vertical angles*. The vertical circles of most theodolites are graduated from 0° to 360° with the 0/360 graduation pointing vertically upwards to the zenith. The circle reading on face left with the telescope horizontal will be 90°. The reading of the circle is best described as a *zenith angle*. The use of the term *vertical angle* should be confined to angles of elevation above the horizontal and angles of depression below the horizontal. In the above case of the telescope being horizontal with a zenith angle of 90° the corresponding vertical angle will be 0°. Figure 2.14 illustrates further the relationship between these two types of angles. Using pocket calculators or computers it is no longer necessary to reduce zenith angles to vertical angles. As long as face left

HORIZONTAL / VERTICAL / ZENITH ANGLES

AT ...*Stn. A*............ FOR ...*Braced   Quadrilateral*.... DATE ...*14/2/88*...
OBSERVER ...*N.R.C.*........ INST. TYPE AND NO. ...*T2 - 14926*... INST. HT. *1.315m*.... ABOVE ...*Nail*...
BOOKER ...*N.R.C.*........ WEATHER ...*Overcast, Cool*........... REMARKS ...*Good Conditions*....

AT	TO	FACE LEFT			FACE RIGHT			MEAN PONTING			ANGLE			TARGET POINT
		DEG	MIN	SEC	DEG	MIN	SEC	DEG	MIN	SEC	DEG	MIN	SEC	
A	B	000	00	12	180	00	16	000	00	14	000	00	00	
✓	C	047	19	24	227	19	30	047	19	27	047	19	13	
✓	D	086	43	32	266	43	40	086	43	36	086	43	22	
A	B	090	21	35	270	21	41	090	21	38	000	00	00	
✓	C	137	40	50	317	40	58	137	40	54	047	19	16	
✓	D	177	04	55	357	05	03	177	04	59	086	43	21	

POINTS TO NOTE :-

1. AT LEAST TWO COMPLETE ROUNDS OF ANGLES OBSERVED
2. FACE LEFT OBSERVATIONS TAKEN CLOCKWISE
   FACE RIGHT OBSERVATIONS TAKEN ANTI-CLOCKWISE
3. DIFFERENT CIRCLE AND MICROMETER SETTINGS SET AT START OF EACH ROUND
4. IF ZENITH OR VERTICAL ANGLES ARE OBSERVED THE EXACT POINT OBSERVED
   TO ON THE TARGET SHOULD BE INDICATED

MEANS :-

∠BAC   047   19   14.5
∠BAD   086   43   21.5

Fig. 2.13 Example of a theodolite booking sheet.

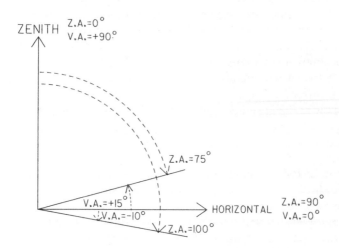

Fig. 2.14 Relationship between vertical and zenith angles.

zenith angles are used the correct signs for differences in height will result when the cosine function is used. On some old optomechanical theodolites the vertical circle is graduated with the 0 graduation corresponding to the horizontal on face left but these are very rare these days.

When zenith angles are being observed on both faces the face right reading should be subtracted from 360° and this result meaned with the face left reading to give a mean zenith angle.

Example:
    Consider the following two readings

Face Left	087°12′14″
Face Right	272°47′40″
360° − FR	= 087°12′20″
Mean zenith angle	= 087°12′17″

It is important to check whether the instrument in use has an altitude bubble to define the horizontal index for the vertical circle or if the instrument has an automatic indexing system. If an altitude bubble is fitted then this has to be centred before *every* reading of the vertical circle. This is similar in practice to using a tilting level as opposed to an automatic level. All electronic theodolites have automatic indexing of the vertical circle.

The most common reason for reading zenith angles when setting out is the determination of slope for correction of slope distances to horizontal and vice versa. Since the angles are read in terms of the horizontal plane through the centre of the vertical circle, it is necessary to observe on a mark at the far point which is either the same height as the instrument or is at least known, so that allowance can be made.

In practice, when measuring the general slope between marks it is usually sufficient to sight on the estimated height of instrument on a ranging pole at the far mark. The height of the instrument can be measured with a similar pole beforehand. No great inaccuracy will occur as the angles are usually small and the sine function changes very slowly between 85° and 95°.

On short distances, under a tape length, it is often expedient to measure the angle direct to the mark and measure the slope distance direct to the trunnion axis. With modern tension grips the tension and sag errors in this form of suspended (catenary) taping are balanced to acceptable accuracy for setting out work. When measuring the distance ensure that the telescope is aligned on the far mark, so that the centre of the trunnion axis forms a mark.

All the foregoing is not to say that there may not be occasions when it is necessary to measure both the height of instrument and the angles with some degree of exactness, but they will be obvious. However, the methods suggested will in normal situations contribute to speed of operation without the sacrifice of adequate accuracy.

There are several suitable ways of measuring the height of the instrument:

• with folding pocket tape
• by measuring against a level staff opposite the trunnion axis
• by marking a ranging pole and measuring later.

Some modern instruments (e.g. Kern) are equipped with plumbing rods (telescopic) which show the height of the instrument above the point as they extend.

Optomechanical Reading Systems

Optomechanical reading systems are of two basic kinds, scale reading and micrometer reading. Additionally the instruments may be what is known as single reading or mean reading. In the former the reading taken after setting the micrometer index is from one side of the plate only. In mean reading types any errors in eccentricity are meaned by the display including a view of graduations from points 180° apart. Scale reading types are also single reading.

Scale Reading

The image of the plate markings seen in the reading eyepiece is sufficiently large to make the position of the reading hairline usually readable by inspection directly down to single minutes of arc and by easy estimation to the nearest 20 or 30 seconds, sometimes down to 10 seconds.

Micrometer Reading

In single reading types the micrometer head must be turned so that the image of the nearest main scale reading moves into line with a setting index. This in turn moves the minutes and seconds scale an amount which is a measure of the odd minutes and seconds in the reading. The movement of the main scale reading when the micrometer head is turned is apparent, not real. The final reading is the sum of the main scale whole degrees and tens or twenties of minutes and the micrometer scale minutes and seconds to whatever degree of fineness it is figured.

In some mean reading types the micrometer head must be moved until coincidence between images from opposite sides of the plate is achieved. The main scale reading at this point is taken and to it is added the reading from the micrometer scale in odd minutes and seconds as before. In other types the micrometer moves a setting index, which is itself a mean, until it straddles the nearest main scale reading to which is added the odd minutes and seconds of the micrometer scale.

Figures 2.15, 2.16 and 2.17 show displays that are representative of the main systems.

(1) The display shown in Fig. 2.15 is from an optical circle 20 second micrometer reading theodolite by Hilger & Watts. The display is viewed through a small reading eyepiece which can be turned to face the observer. It must be focused first to bring the image of the display sharp. Light is reflected into the display by an angled mirror. The arrangement is typical of micrometer optical circle instruments. Both horizontal and vertical scales can be read but the micrometer must be set to the scale in use first by turning the milled head on the right-hand standard.

 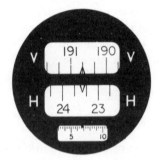

Fig. 2.15

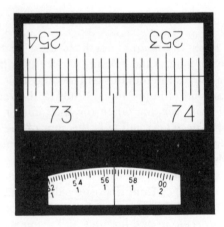

Az: 73° 27' 56,7"

Fig. 2.16

To read:

(i) Turn the micrometer head until a main (H) scale reading exactly bisects the setting index (in example at 23°20'). Turning the head produces apparent movement of the main scale reading and actual movement of the micrometer scale in the small window.

(ii) Read the whole degrees and 20s of minutes (clockwise).

(iii) The micrometer scale now shows the minutes and 20s of seconds (anti-clockwise) opposite its own index mark. Each 5 minute graduation is figured and each minute is divided into three 20 second divisions. In the example for the horizontal scale (H) the reading is 11'40".

(iv) Complete reading is the sum of the two readings, e.g.

Main scale reading	23°20'
Micrometer scale	11'40"
Complete reading	23°31'40"

To read zenith angles, first set the micrometer so that a vertical circle reading bisects the setting index and read as before.
(With acknowledgements to Rank Precision Industries Ltd.)

(2) The display shown in Fig. 2.16 is from the Wild T.3 micrometer mean reading theodolite. The display is viewed through a small eyepiece alongside the main telescope. A changeover knob brings either the horizontal or the vertical circles into view. The micrometer scale can be read directly to single seconds and by estimation to smaller parts.

To read, the inverter knob must first be turned to view the desired circle. The micrometer head is then turned until coincidence between two opposing graduations of the two main scales is achieved. The main scales appear to move in opposite directions when the knob is turned.

The main scale divisions are of 4 minutes and in the example coincidence occurs at a reading of   73°26'
To this must be added, from the micrometer window,   1'56.7"

giving a complete reading of   73°27'56.7"

which is a mean from both sides of the optical circle indices 180° apart.
(With acknowledgements to Wild, Heerbrugg (UK) Ltd.)

(3) The display shown in Fig. 2.17 is from an optical scale reading 30 second theodolite by Vickers Instruments Ltd. It is viewed through a small auxiliary reading microscope alongside the main telescope. This must be focused to bring the display image sharp, and light must be reflected into it from an angled mirror on the standard.

The main scale is divided into whole degrees and is viewed against a reading scale divided into minutes in such a way that 60 minutes exactly

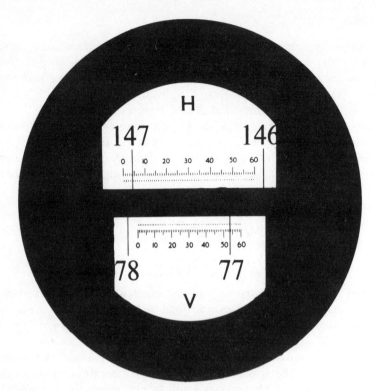

Fig. 2.17

spans the distance between two whole degree marks. To assist in reading, the minutes scale is spread into two layers with whole minute divisions which are staggered by 30 seconds. This arrangement avoids clutter and makes it easier to decide on the seconds portion of the reading where the whole degree mark cuts the scale. A 20-second version would have three layers denoting the 20″ and 40″ positions.

To read:

(i) Note the degree mark which cuts the minute scale, only one can do so except at a whole degree reading.

(ii) Note the number of whole minute divisions from 0 on the minutes scale and decide if the line cuts a whole minute division, a 30 second division or somewhere between. Skilled observers can estimate down to 15 seconds or less.

In the example the horizontal reading is 147°04′15″ and the vertical 77°53′45″.

(With acknowledgements to Vickers Instruments Ltd.)

Testing and Adjustments

There are five important tests on the theodolite which the setting out engineer should be able to do. Of less importance, but nevertheless useful, is the engineer's ability to carry out the relevant adjustments of the theodolite after each test. These tests are:

(1) Plate bubble
(2) Horizontal collimation
(3) Vertical collimation
(4) Trunnion axis
(5) Optical plummet.

Of these, the adjustments for tests (4) and (5) may not be possible since some instruments are not made with the means of adjusting the trunnion axis or with an optical plumbing device.

The Plate Bubble Test

The plate bubble is tested by levelling the theodolite normally, as described previously in the section for setting up the instrument. The bubble is adjusted in its tube until it does not move when the alidade is rotated. The theodolite is now level. If the bubble is not in the centre of the tube it can be brought to the centre by turning the bubble adjusting screws with a tommy bar. This should be rechecked by turning the alidade and readjusting if necessary. This is the simplest adjustment that can be made to a theodolite and probably the most useful.

The Horizontal Collimation Test

A theodolite is correctly collimated in horizontal terms when the vertical plane of the line of sight passes directly through the axis of rotation. In other words, the line of sight is rotating about the same point as the measuring circle or plate.

The test is simple to do, and should be carried out as follows:

(1) Set up and carefully level the instrument so that some sharply defined distant object can be viewed.
(2) Sight on the object at face left and transit the telescope to view an engineer's scale fixed some two or three metres away. It can be taped to a window ledge on the site office or to the instrument case on the ground.
(3) Note the reading of the vertical hair on the scale in whatever units it is divided.
(4) Re-sight the distant object at face right and again transit to view the scale. Note the reading as before.

If the readings on both faces are the same then the instrument is in adjustment in this sense. If they are not, proceed as follows:

(5) Unscrew the cover which protects the reticule adjusting screws.

(6) Re-sight at face right on the scale reading and, using the slow motion screw, traverse the instrument one quarter of the way towards the face left reading.

(7) Transit the telescope and view the distant object, which will now be displaced.

(8) Using the small spanner or tommy bar provided with the instrument, gently slacken one of the horizontal screws holding the reticule frame; note the direction of the movement of the hair line.

(9) With this knowledge adjust both screws so as to move the hair line back to the position where it bisects the distant mark, and at the same time leaves both screws in the same condition of tightness as before.

(10) Re-sight at face left, transit, and intersect the scale as before. If the adjustment has been correctly done, the readings will be identical.

Some delicacy of touch is required and a first attempt may not be successful, but the adjustment is well within the scope of the engineer.

The Vertical Collimation Test

A theodolite is correctly collimated in vertical terms when the horizontal plane of the line of sight through the horizontal hair passes through the axis of rotation of the vertical circle, and is additionally parallel with the plane containing the bubble axis or vertical index compensator and the 90° of the reading circle. Owing to the disposition of the bubble it is more correct to say 'parallel with the plane of the bubble axis'.

To carry out the test, proceed as follows:

(1) Set up and level the instrument in a position from which a distant mark near the level plane can be viewed.

(2) Lay on the mark at face left, centre the bubble (if applicable) and read the vertical scale.

(3) Transit the instrument and repeat the procedure at face right.

(4) If the instrument is in adjustment the reduced angles will be the same. If they are not, set the mean value on the circle with the telescope still laid on the mark, and by using the bubble slow motion bring the reading index to the correct position.

(5) This will displace the bubble from centre and it should now be brought back to the centre of its run with the adjustment screws provided. Repeat the observation of the mark to test the adjustment, which may have to be repeated.

Remember that there are different ways of graduating vertical circles in optomechanical models when reading and deciding on the mean values. The principle, however, is the same for both and no real difficulty occurs. It is not necessary to persevere too much to try to eradicate all errors as, if zenith angles are important, they can be read on both faces and the mean values, which will be free from other than observing errors, accepted.

If the instrument has automatic indexing of the vertical circle then, after steps(1) to (3), the indexing system should be adjusted following the procedures in the manufacturer's handbook.

The Trunnion Axis Test

The theodolite trunnion axis is in correct adjustment when its plane of rotation about the vertical spindle is parallel with the plane of rotation of the plate.

The effect of dislevelment on horizontal angle measurement is negligible, unless objects viewed are at large angles of elevation or depression. Since the site engineer may often set marks in line at considerable angles up or down, it can be important, as the line of sight from maximum elevation to depression will follow an inclined plane and not a vertical one.

To carry out the test, proceed as follows:

(1) Set up the instrument near some tall object which will provide a mark at a high angle of elevation.
(2) Level the plate with some care and sight on the high mark.
(3) Depress the telescope to read an engineer's scale a metre or two away at ground level; note the reading.
(4) Repeat the procedure on the opposite face. If the scale readings are different, set the vertical hair to the mean value and once more view the high mark, which will be displaced from the vertical cross hair.
(5) Bring the object back into position on the vertical hair by raising or lowering one end of the trunnion axis with the adjusting screws provided.
(6) Repeat (3) and (4) as a check on correct adjustment.

The Optical Plummet Test

In general this is not a test which requires to be done frequently as the factory setting of optical plumbing systems is very stable.

The procedure is as follows:

(1) Set up the instrument at any convenient point over a clear cross mark.
(2) Focus the optical plummet and bring the cross hairs directly on to the mark by manipulation of the footscrews, ignoring the level of the plate.

(3) Rotate the instrument through 180° while viewing the mark through the plummet. If it is in correct adjustment, the cross hairs will remain centred. If they wander, the plummet must be corrected by use of the adjusting screws in the mount. These screws may be waxed over.
(4) Correct in planes 90° apart.

If the errors are relatively small and the setting out is not required to high accuracy then it is not necessary to carry out the adjustments as they can be quite tedious. Even if the setting out is required to high accuracy it is still not necessary to carry out the adjustments *as long as* the setting out is carried out on *both* face left and face right and the mean position taken for the point. It is very simple to introduce large maladjustments if the operator is not experienced in carrying out these procedures. Adjustments should only be carried out by experienced personnel and then only if the differences between face left and face right readings are so large as to make the arithmetic tedious. *As long as the instrument is level*, i.e. the plate bubble does not move when the instrument is rotated, then observing on face left and face right and meaning the readings will eliminate errors 2, 3 and 4. If the optical plummet is out of adjustment the instrument is centred over the point if the plummet reticule describes a circle when the alidade is rotated through 360° and the groundmark is exactly at the centre of this circle.

Linear Measurement

Depending on the accuracy required, methods of taking measurements of lengths can vary a great deal. Those generally available to the engineer are:

(1) Direct measurement using tapes.
(2) Indirect measurement using distances in conjunction with angular measurement (e.g. subtense bar).
(3) Stadia tacheometry.
(4) Electromagnetic distance measurement (EDM).

Of the above, the first is the most common for many routine day to day tasks over relatively short distances. EDM is the next most important especially on larger sites where its great speed easily outweighs the initial cost of the equipment.

Measurement Accuracy

It is important to know and work to the required degree of accuracy. The following figures are a good general guide to the accuracy to be expected from various methods:

(1) Taping over the surface and applying only a slope correction: 1 in 1000.
(2) Taping over the surface and applying corrections for tension, temperature and slope: 1 in 10000.
(3) Taping over the surface with tape supported, correctly aligned and applying all corrections: 1 in 30000 (1 in 100000 in special cases).
(4) Indirect measurement using subtense bar: from 1 in 2000 to 1 in 10000.
(5) Stadia tacheometry with theodolite and vertical staff: 1 in 200.
(6) Using EDM. In general these instruments have a constant internal error plus an error proportional to the distance measured. Therefore their accuracy can be said to improve with distance. Most EDMs used on site have an accuracy of ± (5 mm + 5 ppm) i.e. they can measure 1 km to an accuracy of 10 mm or 1 in 100000. (Note that over a distance of 50 m, for instance, the expected accuracy of 5.25 mm is equivalent to only 1 in 9500.) The very latest models appearing on site have precisions of ± (1 mm + 1 ppm) which equates to 1 in 500000 over 1 km. There are available EDMs of even higher precision but these are seldom used for setting out work.

The accuracies quoted for taping (1) to (3) are for steel tapes with a length of 30 m, the most common site tape. If tapes of greater length are used then the accuracy should be better as less tape lengths will be required for each line measured.

In all surveying and setting out operations the method chosen must often be based on the time and equipment available, but *it is as great a mistake to strive for too great an order of accuracy as it is to persist in methods which result in having to repeat measurements.*

Taping

Types of tape

Tapes are made either of steel or synthetic material. Steel tapes are either coated (e.g. black on white) or etched, when the steel is self-coloured with lighter etched graduations. They should have the standard conditions at which they are accurate marked on the first metre (Fig. 2.18).

Synthetic tapes are similar in lengths and markings but are liable to stretch or shrink by quite large amounts although they are less liable to breakage by kinking. In setting out they should be used for rough work only.

Retractable steel rules, very useful for short measurements, are made in lengths of 1, 2, 3 or 5 m.

Tapes are usually housed in a box reel of leather or plastic with a

Fig. 2.18   30 metre steel tape in box reel.

recessed winding handle. Some longer ones (100 m) are mounted in a framed reel with a folding winding handle. The reel is laid aside in use. Such long tapes have occasional use and advantage on large sites, where the initial measure of distances is long.

Measurement Procedures

Measurement with steel tapes is an aspect of site work which is frequently badly done, both from ignorance and lack of skill. The accuracy obtainable from the method can vary between very wide limits, e.g. from about 1 in 1000 to 1 in 10 000. Even the lower limit is often not achieved, through lack of skill and care.

Details of the skills in the use of tapes are set out in the following pages, and are worth study. It is in the exercise of these neglected skills that there lies one of the keys to successful setting out.

Alignment

To achieve good accuracies reasonable care must be taken to keep the tape in the correct line and free from curvature, as well as taking account of the other factors mentioned previously. Lining in by eye with the assistance of ranging poles will generally suffice, but if higher orders of accuracy are required the points at the end of each tape length may require lining in with a theodolite beforehand. However, nowadays, if the distance to be taped is so long that a theodolite has to be used for lining in then it is most likely that EDM would be used to measure the distance.

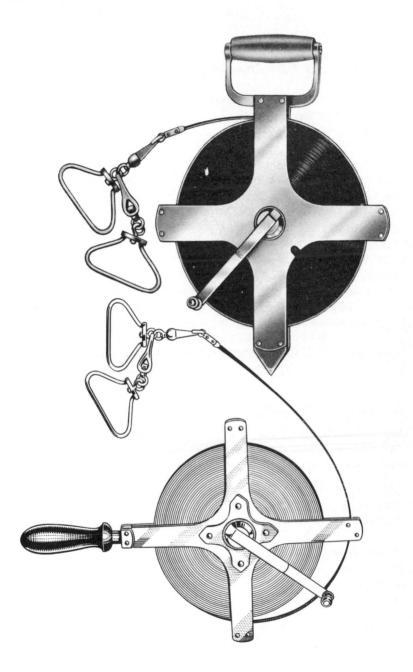

Fig. 2.19 100 metre steel tapes on winding frames.

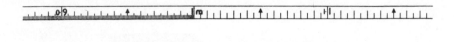

Fig. 2.20 Typical tape graduations.

Measuring Short Distances

The common site method with short tapes, using a nail mark as an anchor at the zero end, may have to be used. If so, the engineer should make sure that the peg is firm and will not be sprung by the tape tension and that he or she knows how much to allow for the loop and the nail, by a check measurement the first time the tape is used. Using tension grips is still an advantage, and when a complete 30 m length is used, some effort should be made to support the centre.

For more accurate short measurements it is better to support the tape if possible and use someone at each end. When using a tape with a loop only it is difficult to hold at zero (the outside of the loop) and better results will be obtained by making a grip (a leather bootlace serves very well), and measuring from the first 100 mm mark on the tape, making the obvious allowance at the other end. The tension grip can be used or a tape grip with a spring balance if the accuracy demands.

Measuring Long Distances

This requires a team effort and clear understanding of the procedure by two people if it is to be done quickly and accurately.

The equipment required is as follows:

- Ranging poles   2 (minimum)
- Steel tape   1
- Marking arrows   10
- Tape grips   }
- Spring balance }   if required for accuracy
- Note-pad

Procedure

(1) Mark ends of line to be measured with ranging poles.
(2) No. 1 holds zero end of tape near first mark.
(3) No. 2 checks that he or she has 10 arrows, and walks towards the far mark counting paces; at 30 paces (for 30 m tape) No. 2 slows down and turns about to reel out the last metre or so of the tape.
(4) No. 2 holds reel clear of the body to one side and responds to lining in signals from No. 1.

(5) No.1 signals to move No. 2's reel hand directly into line with the far mark.
(6) When correctly in line, 1 and 2 together pull on the tape to raise it clear of the ground to straighten it and to clear obstacles.
(7) 1 and 2 lower tape to the ground under tension and 1 checks that 2 is still in line.
(8) No. 1 holds zero point on mark and shouts 'On' (Note 1).
(9) No. 2 tensions tape and drives in first arrow exactly opposite the 30 m graduation (Note 2).
(10) No. 2 then shouts 'Check', to which No. 1 replies 'On' when zero is again on the mark.
(11) No. 2 checks the position of the arrow, adjusts if necessary and indicates satisfaction by releasing tension and standing up.
(12) No. 2 now moves on, counting paces as before and trailing the tape which it is best for No.1 to leave free. Alternatively the tape can be supported clear of the ground during movement.
(13) Steps (3) to (11) are now repeated for the next tape length (Note 3).
(14) When No. 2 reaches the final mark he or she will turn about and reel in until the zero end is near the last arrow.
(15) The final portion of the distance is now noted on the measuring pad (Note 4).
(16) Before No. 1 moves the last arrow a check against gross error is done as follows:

No. 1 counts the arrows picked up on the way (Note 3) and says, for example, 'Seven'. No. 2 checks that he or she has 3 arrows left and says 'Check'. If the total does not check it means that an arrow or a tape length has been dropped, and a check must be made.

Note 1 It is sometimes difficult to hold the zero point exactly opposite the mark when the tape is under tension, and the leather loop aid previously described should be used to give a better hold. As with short distance accurate measure, it is better when using taping arrows to hold the 100 mm graduation opposite the mark. This is more easily done and there is less risk of disturbing the arrow.

Note 2 It is sometimes difficult to drive in an arrow exactly opposite the 30 m graduation because of the nature of the surface. It is permissible to move the tape slightly sideways to overcome this.

If the surface is concrete or will not admit an arrow, the arrow should be laid flat and chocked into position with its point opposite the graduation. Alternatively, a scratch or pencil mark can be made and the arrow left indicating its position. Care should be taken to keep the moving tape clear as the taping team move on.

Sometimes because of vegetation it will not be possible to see the shank of the arrow at ground level. In this case the arrow should be driven in with its ring across the line of the tape and adjusted so that its top portion is exactly under the 30 m graduation. No. 2 should indicate this by shouting 'Top of ring'. No. 1 will then know how to hold the tape when reaching this mark. In all cases when measuring from arrows, care must be taken not to disturb them by holding the tape too close on the final measure.

Note 3 Arrows should have a strip of red or yellow bunting tied to the ring to enable them to be seen more easily. Moving along the line, No. 1 picks up the arrows, but before doing so he or she should be sure that No. 2 is satisfied.

Note 4 In measuring the final portion of the distance it may be necessary to measure down to the nearest millimetre. Tapes may only have millimetre marks on the terminal portions, and in this case, if the final leg is short, the tape can be turned end for end and the distance measured direct, by No. 1 holding the final arrow at the nearest metre graduation and saying, for example, 'Holding at 7 metres'. No. 2 can then read the odd portion at the final mark, the distance of this portion being the *difference* between 7 m and No. 2's reading. Alternatively, No. 2 may hold a whole metre graduation opposite the final mark and No. 1 will read the odd portion opposite the final arrow.

The final gross error check must still be carried out, and it is good practice for No. 2, when using the booking pad, to put a stroke on it every time an arrow is planted, as he or she may accidentally drop one which will render the final check dubious. When the engineer is one of the pair, he should be the No. 2.

Note 5 In any measuring operation it is important that a record is kept. Distances should be transferred to the field book in a form that is identifiable, e.g. either by a sketch or a simple description:

Distance main centre line peg 0 + 10 to FMH 19 centre. 114·26 m.
NW corner No. 86 to building line 35·2 m (right angles to building).

The foregoing may appear to be very detailed for what seems to be a simple operation, but this is deliberate. Skill in linear measurement will only be developed if this type of detail is noted and used.

Taping on Slopes

On smooth, even slopes it is better to tape on the slope and correct for slope afterwards, using the difference in height or the vertical angle of the slope. On rough ground with sharp variations of slope the step method is probably more convenient and as accurate (see Fig. 2.21 (a)).

Quite accurate results can be obtained by the use of ranging poles to mark the ends of the separate lengths measured, keeping the tape horizontal throughout. However, it is not sufficient when doing this to

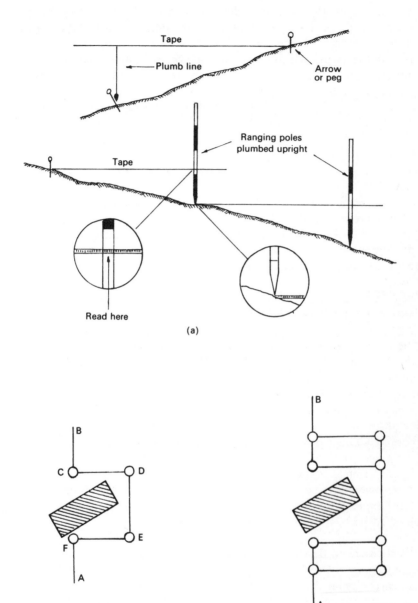

Tape

Plumb line

Arrow
or peg

Ranging poles
plumbed upright

Tape

Read here

(a)

B

C   D

F   E

A

1. Layout careful right
   angles at C D E F
2. Make CD = EF

(b)

B

A

More accurate version
of (b) with more checks
and better control of
line beyond obstruction

(c)

Fig. 2.21   Taping across slopes and round obstacles.

judge the uprightness of the poles. They should be set upright by the use of a small pocket level or the plumb bob and line which the efficient site engineer always carries.

If this is done and the reading marks used are either the point at ground level or the centre of the pole judged visually at points above the ground, very acceptable results are possible. Alternatively, plumb bobs can be used to measure down from horizontal tapes, but the method is slower and a little more exacting.

Measurement of Slope

Slope can be measured by:

(1) straight edge and spirit level and direct measure, e.g. 1 in 4, 1 in 8;
(2) eye level (a large protractor and a plumb line will serve);
(3) ordinary levelling or use of gradient drum on tilting levels;
(4) Abney level;
(5) zenith angle measured with a theodolite.

High Precision Steel Taping

The preceding description of taping procedures will, with care, result in a precision of up to about 1 in 10 000 when all relevant corrections are applied to the measurements. For precisions of better than this then a modified form of procedure is required.

Firstly the line should be subdivided into bays of just less than one tape length or at significant changes in slope. The ends of each bay should be marked with pegs with a nail or pencil mark. The pegs should then be levelled to obtain height differences for each bay for the reduction for slope. Taping should then proceed as described above except that no attempt should be made to hold the tape on an exact division at one end of the tape. When the correct tension has been applied and the tape has steadied with no more pulling to and fro No. 2 should call 'Mark' and *both* ends of the tape are then read simultaneously. The reading from No. 1 is then called down to No. 2 where both readings are recorded. No. 2 then then calls No. 1's reading back as a check. The No. 1 reading is then subtracted from the No. 2 reading to obtain a value for the bay length. The tape is then moved slightly and the process repeated. If the second value for the bay length agrees with the first value to within a predefined limit the two values are meaned and the tape moved onto the next bay. If the two values do not agree the process is repeated a third time and if this third value agrees with one of the first two these two values are meaned and the other value rejected. It is often possible to estimate the readings of the tape to 0.1 mm. It is important that the tape is moved slightly between repeated readings to obtain truly independent readings. Even the best intentioned observer can subconsciously try to obtain the exact same reading as previously. Although this method may seem tedious remarkable accuracies can be achieved. Using a 100 m tape it is possible to obtain accuracies of up to 1 in 100 000 in exceptional circumstances.

Obstructions to Measurement

In measurement over a site there are often obstacles in the way of direct measurement of a distance. There are many ways of overcoming this problem. Two simple ones, not involving the use of a theodolite, are shown in Fig. 2.21 (b) and 2.22.

When a theodolite is available, no particular difficulty arises since a diversion round the obstacle can easily be made either by setting out right angles or a simple direct traverse. A check dimension should, however, always be included and the correctness of the measure not left to depend on angular measure.

Measuring Gaps

If distance measuring equipment of suitable accuracy is not available, gaps which are too large to measure directly can easily be calculated by setting out a small base on one or both sides, observing a braced quadrilateral framework which is self-checking and calculating the side which spans the gap. This is often useful in small bridgeworks.

A somewhat simpler method which will yield good results with care in lining in, is to lay out similar triangles on opposite sides of the line spanning the gap (they should be right-angled) and calculating the gap by proportion against the sides which can be measured (see Fig. 2.22 (b)).

Corrections to Steel Taping

There are a number of corrections which have to be applied to a raw field measurement before it can be used in subsequent calculations. These are:

(1) Tension   too much can stretch the tape, too little will not stretch it enough.

(2) Temperature   steel expands and contracts with changes in temperature thus changing the length of the tape.

(3) Standard   errors in the stated length of the tape because of mishandling or incorrect repairs.

(4) Sag   a tape not fully supported means that the distance taped is curved and not straight.

(5) Slope   measurements are not of the horizontal distance.

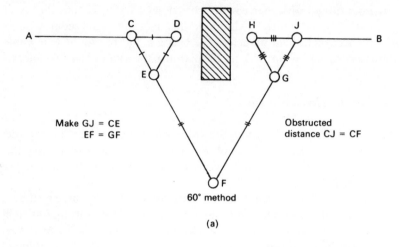

Make GJ = CE
EF = GF

Obstructed
distance CJ = CF

60° method

(a)

Measure AB, AC
and DE

$$\frac{DE}{BD} = \frac{AC}{AB}$$

Measure AB, CD and
all angles.
Adjust angular error
Compute AC from both
base AB and CD.

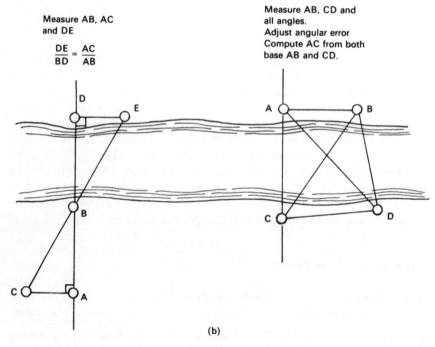

(b)

Fig. 2.22 Taping round obstacles.

For low precision taping normally only the correction for slope has to be applied but if relatively high accuracy is required then all five corrections have to considered and applied *in the correct order as listed above*.

Although made of steel a tape is long and thin and it is quite easy to stretch when pulled from the ends. This stretching can be as much as 15 mm in 30 m. The effect of the stretching can be calculated from:

$$b = \frac{S}{AE} (Tf - Ts)$$

where b is the stretch of the tape
            S is the recorded length
            Tf is the tension applied in the field
            Ts is the standard tension of the tape
            A is the cross-sectional area of the tape
            E is the Young's modulus of the tape.

The standard tension of the tape is almost certain to be 5 kgf with modern tapes. The field tension has to be measured and so a tape grip and spring balance are normally used to measure the force being applied. A simple solution to the problem is to apply the standard tension in the field so that the correction will become zero.

Next the correction for temperature should be applied using the following formula:

$$c = S \times (Ft - St) \times C$$

where c is the correction to the measured length
            S is the recorded length corrected for tension
            Ft is the field temperature of the tape
            St is the standard temperature of the tape
            C is the coefficient of thermal expansion of the tape.

Fig. 2.23   Spring balances for tensioning of tapes.

Fig. 2.24 Tape grip.

Fig. 2.25 Tape clamp.

The standard temperature of a steel tape will usually be 20°C. The field temperature will have to be recorded using a thermometer, preferably one which clips to the tape and directly measures the temperature of the steel itself. If this is not available a normal thermometer should be used to measure the air temperature at the level of the tape. On a cold day on site with the temperature around freezing this correction can be around 10 mm over a 30 m tape length. An alternative to using a steel tape is to use an *invar* band. Invar is an alloy which has a very low coefficient of thermal expansion. If an invar band was used on a freezing day the temperature correction would be only 0.5 mm over 30 m. However invar bands are very rare on construction sites but could be obtained if necessary.

Theoretically a tape is only the length it says it is when the standard tension is applied to it and it is at its standard temperature. Even then age and repairs can produce an incorrect result. Therefore tapes should be regularly calibrated against a known length. Some organisations have a known length permanently marked out but a more common practice is to keep one tape purely for calibration and not to let it be used for site work. This tape can then be taken as the standard and all working tapes checked against it. The resultant correction factor can then be applied to each tape measurement.

Sag errors are difficult to assess, but tension and sag errors (positive) tend to compensate. Sag can be eliminated by supporting the tape. If the tape is not supported throughout its length, errors will occur from sagging of the unsupported part. This can be allowed for in accurate work by using a correction $\dfrac{W^2L^3}{24T^2}$, which is always negative, where:

W = weight per unit length kg/m
L = catenary span m
T = tension applied kgf (normally 5 kgf).

Slope errors can be greater than realised: a difference in height of 0.5 m in the ends of a 30 m tape produces a positive error of 5 mm.

Since all data are shown in true plan it follows that measurement made on the ground must be reduced to the horizontal. The correction can be made either by measuring the angle of slope, or by levelling to each change of slope.

The corrections are:

$$\text{Horizontal Distance} = \text{Slope Distance} \times \sin Z$$

or

$$\text{Horizontal Distance} = \text{Slope Distance} - \frac{\text{Diff Height}^2}{2 \times \text{Slope Distance}}$$

A simple trick on site is to combine tension, standard and sag in a single correction factor. Sag can be reduced by applying a much higher tension in the field than the standard tension. The tape is then standardised with the higher tension applied to the working tape and the standard tension applied to the calibration tape. If the resulting correction factor is applied in all subsequent fieldwork then only temperature, standard and slope corrections need to be applied.

The Care and Use of Tapes

(1) Always keep tapes reeled up when not in use.
(2) Clean and wipe off with a dry rag after use. For etched tapes use an oily rag after wiping clean and dry. Wash metal woven and synthetic tapes and leave looped to dry before reeling up.

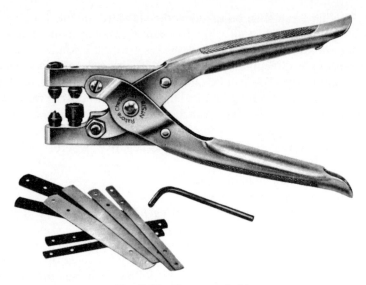

Fig. 2.26 Tape repair kit.

(3) Never pull on a steel tape to straighten a loop; it will always kink and break. Never allow a vehicle to run over a tape even when laid flat and slack; it will almost certainly break.
(4) If right-handed, hold the case in the left hand when reeling out and in, allowing the tape to run between first and second fingers when reeling in. This will prevent kinks and jamming with synthetic tapes.
(5) Run steel tapes straight out from the reel and always reel in when a shorter distance follows a longer one.
(6) Pull on the tape, not the reel; use tape grips if available.
(7) Keep the tape straight and free from sag when in use.
(8) Check against standard from time to time and *always* after repair.

Subtense Bar

This is an indirect form of linear measurement in that the quantity measured is, in fact, an angle and the required distance is derived from this angle. A typical subtense bar consists of an aluminium tube, often in two sections, which fits a standard theodolite tribrach. Mounted at each end of the bar are two targets which are held a constant distance apart by invar wires so that their separation does not change significantly with temperature. The bar is levelled and centred over the groundmark using the tribrach as for a theodolite. A small sighting telescope mounted normal to the bar is used to sight the far end of the line to be determined. This positions the bar perpendicular to the required line.

A theodolite is set up over the other end of the line and the two targets on the bar sighted in turn. Thus the angle *subtended* at the theodolite by the two targets can be determined. For precise results a 1″ theodolite should be used and up to 10 complete rounds of angles should be measured so that a very reliable estimate of the subtended angle may be found.

If b is the separation of the two targets and θ is the measured angle subtended at the theodolite, the distance D from the theodolite to the bar can be calculated:

$$D = \frac{b}{2} \times \cot\frac{\theta}{2}$$

As b is 2 m for most subtense bars this can be simplified to

$$D = \cot\frac{\theta}{2}$$

As θ is the *horizontal* angle subtended at the theodolite then the distance D is the *horizontal* distance to the centre of the bar and so no corrections need to be applied to this distance.

The subtense bar is a very useful method for determining distances across obstacles which cannot be taped and if EDM is not readily available. However subtense bars are becoming very rare on construction sites these days but it is possible to fabricate one using a timber batten with suitable targets marked or fixed at each end (see later, Fig. 2.37). This can be mounted on a tripod head and, by sighting along a set square, positioned perpendicular to the line to be measured. The distance between the two targets should be measured by pocket tape after construction. Alternatively a level staff placed horizontally on a tripod head can be used with suitable graduations used as the targets. Some reduction in accuracy will have to be accepted using these alternatives but this may be acceptable in particular circumstances.

Stadia Tacheometry

Stadia tacheometry, once a very common technique for collecting detail information on site, is now virtually extinct. On all but very small sites EDM is far more efficient and significantly more accurate. For the very rare occasions when stadia tacheometry may be employed the relevant formulae are:

$$D = (u - l).K.\sin^2 Z$$

$$\text{and } RL_s = (u - l).K.\sin Z.\cos Z + RL_i + h_i - m$$

where u = the upper stadia hair reading
    l = the lower stadia hair reading
    m = the middle (horizontal) cross hair reading
    Z = the zenith angle
    K = the stadia constant (almost certainly 100)
    D = horizontal distance from instrument to staff
    RL_i = reduced level of instrument station
    RL_s = reduced level of staff point
    h_i = instrument height above groundmark.

Electromagnetic Distance Measurement (EDM)

Since the advent of micro-wave distance measuring equipment, a development of radar, shortly after World War II, the parallel development in solid state devices and electronics has put the principle and the facility into the field of the setting out engineer whose province is measurement.

In some ways, site measurement is often more difficult than original survey measurement since, while the surveyor works generally untrammelled by anything other than the difficulties of the terrain, the setting out engineer must do a great deal of his measurement amid the hurly-burly of the construction contract.

The ability to measure distance accurately and quickly which is afforded by modern EDM equipment especially developed with the construction contract in mind, is to be valued not only for its accuracy but for its speed, with all the benefits associated with freedom from expensive errors, delays, etc., quite apart from the economy in the engineer's time. No cost-conscious engineer on medium-sized and large contracts can afford to neglect the benefits which the use of EDM equipment brings: it is generally so reliable in operation, and now much less expensive than the original micro-wave equipment, that its use for all large contracts should be mandatory.

EDM is another indirect form of linear measurement in that the instrument itself actually measures the phase difference between the transmitted signal and the signal received back from the far end of the line. Since the wavelength of the signal is known, the distance is directly related to the number of full wavelengths traversed and the proportion of the phase shift. The number of full wavelengths traversed is capable of being resolved by the equipment by variation of the modulating frequencies. There are minor corrections to be made since the wavelength of the radiation is dependent upon the temperature, pressure and relative humidity of the air it passes through. These variables can be measured and corrections easily applied.

The original equipment (the Tellurometer) used two stations, a master and slave, and was capable of measuring distances of the order of 50 km to an accuracy of $\pm$ 15 mm $\pm$ 3 or 4 parts per million. The development of equipment for shorter distances (up to 10 km) has made use of the infra-red and visible light parts of the spectrum rather than the radio portion. This has distinct advantages in that it is somewhat easier to concentrate the beam width in the optical mode, and use can be made of simple reflecting prisms.

A number of instruments use a pulsed laser signal in which the time interval between the transmission and reception of the signal is measured and converted into a corresponding distance based on the velocity of the signal. At ranges of a few hundred metres or so it is possible to bounce the laser pulse back off objects without the need for reflecting prisms.

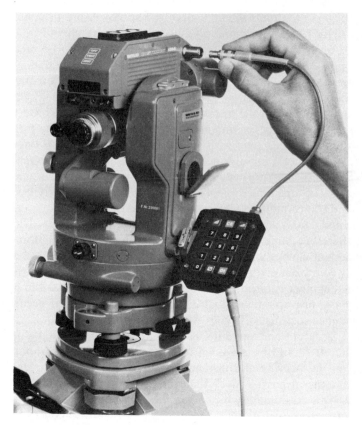

Fig. 2.27 Wild D1−4 EDM fitted to a Wild T1 theodolite.

Advances in electronic circuitry have also made it possible to present the measurement in digital read-out form on a display of the type familiar to all users of electronic calculating machines (originally it was necessary to read a dial or dials and do small calculations). Additionally, the physical size of the equipment has been much reduced, as has its power consumption. Currently there are a number of models capable of being mounted on the telescope of a theodolite, with either built-in data correction facilities or separately attachable calculation modules. These allow for observations to be read out directly as horizontal distances or vertical intervals. For the large contract they offer the setting out engineer greatly increased opportunities for quick and accurate work, particularly on the difficult sites often encountered on large engineering works. Most instruments have 'tracking' capability, i.e. they permit small changes of distance to be recorded so that an engineer can direct the movement of a reflector towards or away from the instrument to obtain a desired distance. In use a reflecting prism (or prisms) must be mounted at the distant point. For rough work this can be a single pole (though with some means of ensuring its verticality—a plumb rod or a circular bubble). More usually an instrument tripod capable of being plumbed over a mark should be used. The number of prisms required depends on the distance to be measured. A single prism usually suffices up to about 800 m, beyond this an array may have to be used, the number of prisms depending on the make of instrument.

When setting out exact distances it is usually quicker to have the tripod positioned accurately for line and roughly for distance. The fine measure for distance can be made with a tape and a peg driven on line and then accurately checked for distance from the plumb line with the tape. The tripod remains undisturbed, perhaps for other local measurement.

The use of such instruments for setting out by what is essentially a bearing distance method while offering increased accuracy and economies in time and labour also offers increased opportunity for error. The engineer must therefore build in to the plan of setting out a suitable number of independent checks which will ensure that the results are 'locked in' and sensibly free from error. There is much scope for ingenuity in this respect and the engineer is afforded the same facility for checks as the sophistication of the instrument originally affords in the setting of accurate angle and distance measures. It goes without saying that care and skill are still needed, accurate and versatile instrumentation will not make up for lack of either. It should be a golden rule that no major mark should depend solely on the bearing and distance from a single station for its position.

The merits of having more than one prism tripod should not be overlooked. This equipment which so facilitates site measurement up to hundreds of metres over difficult terrain without regard to construction work going on (except for interruption to the line of sight), can do in an hour what may take a day by ordinary taping methods. It should not be slowed down for lack of supporting prism arrays.

Calibration of EDM

Modern EDM can produce very precise results very quickly but it is important that the engineer does not take the displayed distance for granted. EDM, like all other instruments, needs checking and adjusting at regular intervals if the specified accuracy of the instrument is to be met. The sections which follow are applicable to infra red, visible light and laser EDMs, including those which are incorporated in electronic tacheometers.

There are five main sources of error which have to be considered when checking an EDM:

(1) Lack of collimation
(2) System zero error
(3) Cyclic error
(4) Scale error
(5) Atmospheric effects.

Lack of Collimation

If the centre of the EDM beam is not pointing directly at the centre of the prism then, because of possible inhomogeneities in the wavefront of the signal, errors can result in the measured distance. Therefore it is important that the instrument is lined up on the prism as accurately as possible. If the EDM is mounted on top of the theodolite telescope then, after being mounted and dismounted a number of times, the theodolite telescope and EDM beam can become misaligned. Even EDMs which are coaxial with the theodolite telescope, as in some electronic tacheometers, can go out of collimation after a period of use.

A simple test for this lack of collimation is to first sight the target on the prism with the theodolite telescope. Most EDMs should have a meter which can be used to display the strength of the signal being received back from the reflector. Using the tangent screws of the theodolite the telescope should be moved left and right and up and down around the target. If the received signal strength *increases* as the telescope is moved *off* the target there is a misalignment of the telescope axis and the EDM axis. The mean of a series of distance measurements taken when the signal strength is at a maximum can be compared with the mean of a series of measurements taken when the telescope is pointed at the target. If there is a significant difference between these two means then the collimation of the EDM will have to be corrected. (This discrepancy can be of the order of 10 mm with some infra red EDMs.) The adjustment procedure is relatively simple if somewhat tedious at times. For top mounted EDMs the attachment block on the theodolite should have two adjusting screws. Some coaxial EDMs will also have adjusting screws. Align the theodolite telescope on the correct point on the target and then, using the two adjusting screws, obtain a maximum indication on the signal strength meter. Check that this is actually the maximum by repeating the test with the theodolite tangent screws described above. If necessary use the adjusting screws to improve the alignment. Repeat this procedure until a maximum signal strength is indicated when the theodolite is correctly sighted on the target. If there are no adjusting screws on the instrument and the discrepancy found in the above test is significant then the instrument should be returned to the manufacturer for servicing.

System Zero Error

Because of the need to balance an instrument correctly it may be necessary for the manufacturer to place the measuring centre of the instrument in front of or behind the centre of the mount. The reflector may also be constructed in this manner. The amount by which the total system—instrument plus reflector—varies from the positions of the centre of the mounts is known as the *system zero error*. The measurement then has to be corrected by the system zero correction, S_o, also known as the prism constant or offset, which has the same magnitude as the system zero error but the opposite sign. Figure 2.28 shows the situation where D_c is the slope distance between the centres of the two mounts.

$$D_c = D_m + S_o$$

Some manufacturers make the offset in the prism cancel out the offset in the instrument resulting in an S_o of zero. Other manufacturers do not. It is therefore very important to check which case applies with the equipment being used as S_o can be of the order of 100 mm in some instruments. A common situation on site is when a new EDM is purchased to replace an obsolete EDM but the prisms for the old instrument are retained. This is because prisms are expensive and can be used for many years before the coating on the glass begins to deteriorate. Therefore a system zero correction for this combination will have to be determined. If there is any doubt at all about what the actual system zero correction is then a very simple test can be carried out to determine it.

A minimum of three points should be set out in a straight line with tripods and tribrachs centred over them. The EDM is set up on A (Fig. 2.29) and the prism on B and the distance AB measured. The prism is

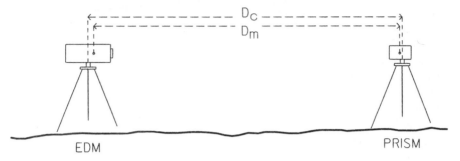

Fig. 2.28 System zero error in EDM-prism combination.

Fig. 2.29 Typical baseline layout for system zero calibration.

then moved to C and AC measured. The EDM is moved to B and the distance BC measured. The system zero correction, S_o, is found by:

$$S_o = -(AB + BC - AC)$$

A better estimate of S_o can be found by using more points on the line and either finding values for S_o for a series of combinations of distances similar to above and taking a mean value or, preferably, computing a least squares estimate of S_o from all the measurements. Note that in these tests the actual distances between the points does not need to be known as long as the tripods and tribrachs are not disturbed as the equipment is being moved around.

Having obtained a value for S_o then the measured distance has to be corrected by this amount. Most instruments allow a value to be entered into their microprocessors and then apply the correction automatically but other instruments do not have this facility and the correction must be applied manually later. When setting out with EDM allowances must be made for this correction in the computed setting out distance.

Another large error which can be introduced in top mounted EDMs is illustrated in Fig. 2.30. Because of the separation, s, between the trunnion axis of the theodolite and the centre of the EDM, as the theodolite telescope is elevated the centre of the EDM moves back and as the theodolite telescope is depressed the centre of the EDM moves forward. In some systems the prism has a similar separation, s, above its axis of rotation so that the two movements will compensate. However many other prism systems rotate around the centre of the prism. Therefore the

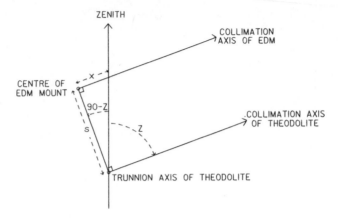

Fig. 2.30 Effect of zenith angle on measured slope distance.

measured slope distance will have an error, x, in it. This can be corrected for by using the formula:

$$x = s. \tan (90 - Z)$$

This error, x, can be of great significance but is often not considered. Obviously as the telescope is rotated further from the horizontal x will increase but even at only small deviations from the horizontal x can be quite large. In some older more bulky theodolite EDMs the separation, s, is about 100 mm. With a zenith angle of only 80° then the error x is 17 mm. With a zenith angle of 60° x is 58 mm! Even with more modern compact EDMs which can fit between the standards of the theodolite x can be significant. Therefore before using an EDM the separation, s, should be measured and values for x calculated for representative zenith angles to see if the error is significant or not. It should *never* be ignored! Note that in virtually all EDMs it is not possible to automatically apply a correction for this error so great care must be taken when measuring or setting out using inclined lines of sight.

Cyclic Errors

Cyclic errors are caused by contamination between the transmitted signal and the received signal. The effect of this contamination varies with the phase difference between the two signals producing errors which follow a cyclic pattern. These errors can be up to 10 mm or so with some equipment although they are normally less than the specified accuracy of the EDM and can then be ignored for most purposes. However if high accuracies are required then these errors should be tested for and corrections applied. This is normally a laboratory test and a typical method is detailed in Burnside[1].

Scale Errors

Scale errors are introduced when the frequencies of the signals being generated are not the same as the values programmed in the instrument's microprocessor. These errors are seldom significant in infra red EDM unless the instrument has not been regularly serviced and the crystals which generate the signals have begun to deteriorate. Again this can only be tested for in a laboratory and should be carried out by the manufacturer during routine servicing.

Atmospheric Effects

Most EDMs assume a value for the wavelength of the carrier and measuring signals. Similarly instruments using pulsed measuring techniques assume a value for the velocity of the signal. However changing atmospheric conditions, pressure, temperature and humidity will affect the velocity, and hence the wavelength, of the signal thus producing an error in the deduced distance. The effects of these atmospheric variables can be calculated and corrected. Most modern EDMs have a facility for inputting a correction for these effects, normally expressed as parts per million (ppm), which can be calculated from supplied formulae or read off supplied graphs in the instrument handbook. Therefore, for good results, equipment for measuring temperature and pressure should be available on site. It is not possible to completely correct for atmospheric effects as the atmosphere varies along the line being measured so that there will always be some residual error.

Summary of EDM Errors

The major source of error in EDM measurements is normally caused by ignoring or applying an incorrect system zero correction. The vertical separation of telescope and EDM can also produce very large errors if not corrected for. Lack of collimation can also produce significant errors as can ignoring atmospheric effects. Cyclic errors and scale errors can usually be ignored as long as the instrument is regularly serviced. Kennie et al.[2] provides a useful list of EDM calibration facilities in the UK.

Lasers

'Laser', like 'radar', is derived from an early description: 'Light Amplification by the Stimulated Emission of Radiation'.

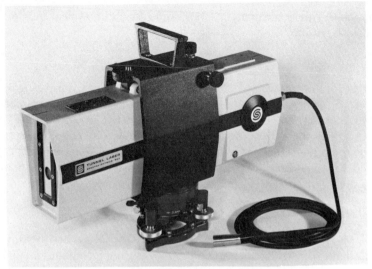

Fig. 2.31   Spectraphysics tunnel laser.

The characteristics of laser light are remarkable in that it is of a single frequency, the beam is coherent and very nearly parallel. This makes it possible to project a very thin beam of intensely bright light over quite considerable distances from the small equipment developed as construction and surveying aids. The technology of the laser has already made great advances and various types of outstanding power and usefulness for many purposes are already in existence.

The present main advantage of the laser to the construction engineer is that it provides a visible line of sight or reference line which can be referred to when needed and presents no obstruction to other activity at other times.

Small building lasers are available which project this beam of light in a horizontal or vertical plane according to the setting of the source. To the operative this can be a stringline for both line and level or may serve simply as a datum.

Other types can be fixed to theodolites for the projection of centre lines. One such uses a light guide to project the beam through the optical axis of the theodolite telescope, while at the same time allowing the instrument to be sighted normally. All that is required is to fix the special eyepiece to the instrument and connect to the laser by an optical fibre type of flexible light-guide of pencil thickness.

The usefulness of lasers in underground work for tunnel centre lines is obvious, and equipment is available for permanent installation in headings with suitable adjustments to permit correct alignment.

It is very obvious that to be useful the source must be stable and capable of fine adjustment for line and level. If the beam can be focused

Fig. 2.32 Spectraphysics rotating beam laser.

on a distant accurate mark it is easier to set up than by adjustment which relies on level bubbles. If the line is to rely for level on spirit bubbles, the accuracy obtainable is restricted to the accuracy of the bubble at the distance to which the beam is to be used as a reference. A 20-second bubble (normal plate bubble on a level or theodolite) gives that accuracy (to ± 20 seconds of arc). More sensitive bubbles tend to be large and vulnerable.

However, recent models have built-in compensators similar to those in automatic levels, which, provided that the instrument is within a fairly broad tolerance for level, automatically project the beam in a level plane to within finer limits than the bubble capabilities.

A type which projects a level beam continuously rotated through 360° automatically shuts off when out of level. On sites where a large number of levels has to be set this type of equipment is most useful as it allows operatives to intercept the beam over a wide area for levels thus freeing the engineer for other work. A level stave has been developed for use in this connection having a beam-seeking traveller which runs up or down until it centres on the point at which the laser beam cuts the staff. With a schedule of settings provided the engineer operatives engaged in various construction tasks can check their own construction levels. While not freeing the engineer from ultimate responsibility for setting up and checks great savings in time and materials can result from the use of such devices − more than enough to offset the cost of the equipment on any but the smallest contracts.

A laser set up as a reference will need to be checked from time to time against disturbance from whatever source, e.g. accident, vibration, temperature change, etc. In use the beam is referred to by intercepting it on any stable surface from which measurement can be done. Targets with concentric marks on the central right-angled cross allow the target to be accurately positioned in relation to the beam centre despite any diffusion or broadening with distance. Close to, the beam itself may show a sufficiently sharp spot of light to constitute an adequate mark.

There is little doubt that the costs can be more than recouped on any but the smaller contracts. On tunnelling contracts or large contracts for drainage or multiple footings for high density housing the cost would be favourable. It is for the engineer to decide whether the time/cost factors are sufficient to justify use.

Electronic Tacheometers

A type of instrument becoming more and more common on site is the electronic tacheometer, often known as a 'total station'. This type of instrument consists of an electronic theodolite combined with an EDM, usually with additional facilities for reducing and recording measurements and results. Some instruments are single units with the EDM integral with the theodolite whereas others are modular, i.e. the theodolite and EDM are available as separate independent units but can be connected together to form a single operating unit.

Most instruments have a small microprocessor for simple calculations such as reducing measured slope distances to horizontal distances and differences in height. More elaborate instruments can also compute coordinates from the measurements if certain information is input through a keyboard (Fig. 2.33). More sophisticated computations can be carried out by adding optional processing units as with the Zeiss Elta 2 (Fig. 2.34) or by interfacing the instrument to pocket calculators or portable computers. These instruments give the engineer very powerful facilities available on the tripod.

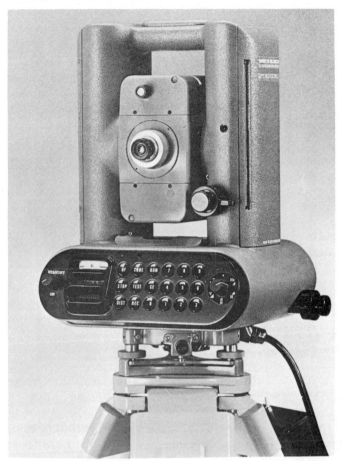

Fig. 2.33 Wild TC1 electronic tacheometer with reducing and recording capabilities.

Electronic Fieldbooks

'Electronic fieldbook' is a widely used term to describe a whole series of devices for recording data electronically on site. Three different recording media are commonly used — magnetic tape cassettes, solid state memory units and 'bubble' memory units. Early recording systems used punched paper tape but these were very bulky and the paper tape very fragile and this medium has virtually disappeared from use. Most systems first record the data and then read them back for verification.

All three types of recording media in use have advantages and disadvantages. Magnetic tape cassette drives are relatively large and are susceptible to dust and moisture entering through the cassette cover. The

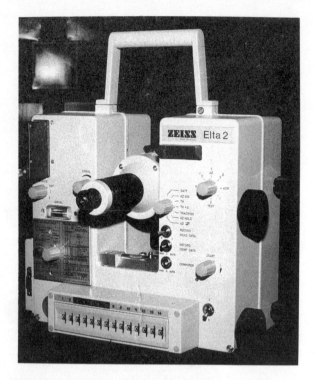

Fig. 2.34   Zeiss Elta 2 electronic tacheometer with automatic recording unit and
computing module.

cassettes themselves are relatively fragile and the recording and verification process slow. Their main advantages are that the cassettes are cheap
and have a large capacity. Additionally when a cassette is full a new blank
cassette can be inserted and work can continue. Solid state memory units
are the most common systems. Recording and verification is rapid, the
memory modules are robust but have limited capacity and are relatively
expensive. 'Bubble' memories are the most advanced and the most expensive but have large capacities.

Some systems can be used as separate units with data being keyed in
manually for storage. Most can be interfaced with EDMs, electronic theodolites and electronic tacheometers for automatic recording. Some can be
loaded in advance with, say, previously computed setting-out data, and
the information can be recalled in the field for reference. Other units are
either supplied with a series of standard surveying programs or can be
programmed as desired by the operator.

Memory units can be connected to computers, either in the site office
or via a telephone link, and the recorded data downloaded for subsequent
processing.

Fig. 2.35   Kern E2 electronic theodolite (photograph courtesy of Kern Instru-
(Left)   ments Ltd).

Fig. 2.36   Kern modular electronic tacheometer system incorporating E2 electronic
(Right)   theodolite, DM503 EDM, DIF41 interface and Hewlett Packard HP41C
                calculator as electronic fieldbook and field computer (photograph cour-
                tesy of Kern Instruments Ltd).

As with all modern aids electronic notebooks are designed to make
more efficient use of the engineer's time on site. If used in conjunction
with an electronic tacheometer, for instance, vast quantities of data can
be collected quickly and processed rapidly. This subsequent processing is
often done by people not directly connected with the data collection and
many kilometres away at the end of a telephone link. This can introduce
a whole set of new problems for the engineer.

An experienced surveyor or engineer using well-proven traditional
techniques of observing and recording manually has many in-built checks
on the data being collected. Duplicate rounds of angle, if reduced in the

field, will rapidly disclose mistakes in observing or booking and indicate if the necessary precision is being achieved. Simple checks can then be carried out to detect misclosures in triangles or traverses. However if the information is automatically recorded and processed later then mistakes in numbering or observing may not become apparent until much later. In large surveys with modern software packages an error at one station may be spread around all the stations in the network and become very difficult to track down. In manual recording of stadia tacheometry, for instance, a sketch is normally produced with each surveyed point marked and annotated on it. In many electronic methods detail points are allocated specific numerical codes which have to be keyed into the recording unit before each measurement is made. If the wrong code is keyed in this may not be discovered until much later and may necessitate a repeat visit to the site to sort out the mistake. Therefore, in some ways, the engineer's life has been made harder. Being reduced to a mere 'button pusher', as some engineers complain, can actually reduce an engineer's edge and lead to slackness and mistakes. The good engineer should realise that the simplification in operation of modern equipment will enable the engineer to concentrate more on *what* data are being collected and that they are relevant data. In many organisations it is now the assistant who points the instrument and pushes the buttons and it is the engineer who goes around with the prism and detail pole selecting the important features that are to be recorded.

Aids to Measurement and Setting Out

The setting out engineer quickly discovers that measurement, angular or linear, can be exacting and sometimes laborious in the extreme and, as he becomes more experienced, he learns to use aids and methods which help greatly to offset this aspect of the work. The suggestions in the following paragraphs will, it is hoped, assist in the overcoming of simple problems, the solution of which may not be immediately obvious.

Line Marking

There is a constant need to refer to a reference object or line marker when angular observation is being done. Sometimes the mark is near and easily visible (a nail in the top of a peg, for example). At other times it may be obscured by obstacles or may need picking out by some easily identified object such as a ranging pole.

A good line marker, superior to a ranging pole at near and medium distance, can be made from an ordinary building lath (about 50 mm wide) with a black line scribed down the centre of its length. In use it can be

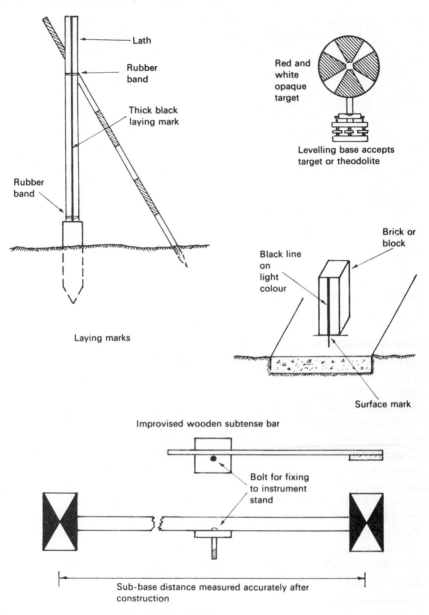

Lath

Rubber band

Thick black laying mark

Rubber band

Red and white opaque target

Levelling base accepts target or theodolite

Brick or block

Black line on light colour

Surface mark

Laying marks

Improvised wooden subtense bar

Bolt for fixing to instrument stand

Sub-base distance measured accurately after construction

Fig. 2.37 Aids to measurement.

placed on a nail mark, held securely by a rubber band, plumbed, and held upright by a ranging pole at an angle. This makes an easily identifiable fine mark, which can remain in position all day without the need for repeated indication of the mark by an assistant.

On surface marks, a tripod to hold a ranging pole upright is sometimes adequate, provided that the point can be seen or the pole is sufficiently distant to constitute a good laying mark.

Another device is an ordinary brick or concrete block with a yellow stripe (paint or marker crayon) on which a black line is scribed. This will sit over the exact position of a surface mark and give an excellent laying mark at short or medium distances.

At the top end of this scale is, of course, traverse equipment, almost mandatory on large contracts. With this equipment, targets, distance measuring prisms or subtense bars can be accurately positioned over marks. If necessary they can be illuminated at night, or in poor conditions. Additionally, of course, if they form part of a set with the engineer's theodolite, time will be saved on those occasions when only the instrument head needs to be removed. There is the added advantage that the mark is central from all points of observation, which, in general, is not true of the other devices.

Linear Measurement

Frequently too little attention is paid to aids in this aspect of setting out, often amounting to the mere provision of a tape without any ancillary equipment, even on contracts where a great deal of important measuring is to be done (see section on tapes, page 47, for useful equipment). From time to time special conditions arise in which the following may be found useful − for example, where a fairly exact measurement must be done.

For short distances (under 100 m or so) old table knives instead of arrows constitute a much improved index mark.

For longer distances, such as the base line of an overland tunnel survey or on a large industrial complex, setting out short pegs on line at multiples of a tape length and capping each of them with a small metal plate, about 100 mm square, provides a surface on which fine marks can be scribed at the exact tape lengths. Levels taken to the tops will ensure that slope is correctly accounted for and it is possible to arrange a set of supports to go forward with the taping so that sag is eliminated. With a correctly tensioned tape (temperature corrected if necessary), very high orders of accuracy can be obtained in this relatively simple way when occasion demands.

At the other end of the scale, when surface conditions are difficult (swamps or rough terrain) a subtense bar 2 m long can easily be made up from ordinary 50 × 25 timber with simple targets at the exact distance required (wood has a very small coefficient of linear expansion). It can be set horizontal on a spare instrument tripod and square to the line of sight by sighting along a set square. It has been used with advantage on, for example, borehole sitings, and over-water measure. As with EDM prisms, it is convenient to do the final measure in the vicinity of the tripod by tape or rule, depending on the accuracy required. A metal levelling staff (not telescopic) can be used in a similar way, extempore.

Last, but not least, is the importance of having an assistant who can be relied upon and can be taught the niceties and requirements of all types of measurement in which he is involved. The engineer should be at pains to show and explain exactly what is required and should give some thought to continuity of employment of the assistant in this important role.

Electronic Aids

With the advent of EDM radial setting out can now be done quickly and accurately over long distances. This introduces problems of communication between the engineer and the assistant. The traditionally used handsignals may not be visible or clearly understood over these distances. One solution is to issue the setting out team with radios and this is quite a common practice. However carrying out fine adjustments with an instrument, trying to talk on the radio and holding a field sheet all at the same time can be quite a task. There is also the major problem of finding a clear radio channel in today's crowded airwaves. Recent developments in electronics, however, now make it possible to pass unambiguous signals over long distances without the need for radios.

One very simple system incorporated in some modern instruments is a light beam shining through a series of coloured filters mounted on top of the instrument. This enables the assistant to find the required line very quickly as the colour of light visible from the assistant's position will change depending on what side of the line the assistant is positioned. The light should be white when the assistant is exactly on the line.

Another, more sophisticated, approach is to incorporate a unit to modulate an EDM's carrier wave signal to carry verbal instructions from the operator. A small demodulator and speaker is mounted on the side of the prism. Thus the operator can speak to the prism holder although the prism holder cannot talk back.

It is also possible, in some electronic tacheometers, to transmit computed information along the EDM beam which can then be displayed on a unit attached to the side of the prism (Fig. 2.38). This makes it possible for the instrument, when in tracking mode, to continuously compute, for example, the coordinates of the prism position. With this displayed on the side of the prism the prism holder then moves until the required coordinates for the setting out point have been obtained. This system also incorporates an audible signal to inform the prism holder when he or she is in the beam.

Fig. 2.38 Kern RD10 remote prism display (photograph courtesy of Kern
Instruments Ltd).

References

(1) Burnside, C.D. (1982) *Electromagnetic Distance Measurement*, 2nd
edn. London: Granada.
(2) Kennie, T.J.M., Brunton, S., Penfold, A. and Williams, D. (1988)
'EDM instruments: Calibration methods and facilities in the UK'. In
Land and Minerals Surveying, **6**, 1, pp. 27–35.

3 Survey Methods and Procedures

On large contracts the engineer engaged in setting out may only be concerned with this one particular aspect of the surveying for the job. However a good understanding of how the setting out data were produced is necessary if the setting out is to be done to the required precision. On small sites the setting out engineer may be responsible for adding more control points or extending the detail survey. Therefore this chapter covers a number of standard techniques for control and detail surveys.

Coordinate Systems and Datums

It is of the utmost importance for the engineer to ascertain what coordinate system or systems are being used on site and to check that all the data have been calculated on the relevant systems. Possible coordinate systems or grids, as they are commonly called, that are often used are:

Site grid – on large schemes imposed at design stage or by engineer for setting out.

Survey grid – used in original survey for production of the site plan.

Structural grid – applied to separate structures to locate the position of structural elements.

The site grid is the one most applicable to general site setting out and may be identical with the survey grid. It is seldom, however, that the orientation of the survey grid is the most convenient for setting out, and a site grid is to be preferred as it will be related to the disposition of the works in a more convenient form for design or setting out.

It is not necessary for the orientation to bear any finite relationship to true or National Grid North, though, as on any site plan, this relationship should be known to within, say, 30 minutes of arc. The coordinates also need not bear any pre-determined relationship to either National or survey grid coordinates. When setting out the grid on site, however, it will, in general, be done in terms of the original survey marks and a relationship will emerge.

81

It is important that the origin shall be sufficiently West and South of all works to render all coordinates positive (i.e. East and North of origin).

It may on occasion be convenient and of advantage to calculate site grid coordinates in terms of another grid; the formula for such an operation is simple and is given below.

$$E = b + E_1 \cos\Delta - N_1 \sin\Delta$$
$$N = a + N_1 \cos\Delta + E_1 \sin\Delta$$

On most sites the grids used are assumed to be plane rectangular systems. However on large sites, such as major highway projects, it is common to work on the Ordnance Survey National Grid. This is a map

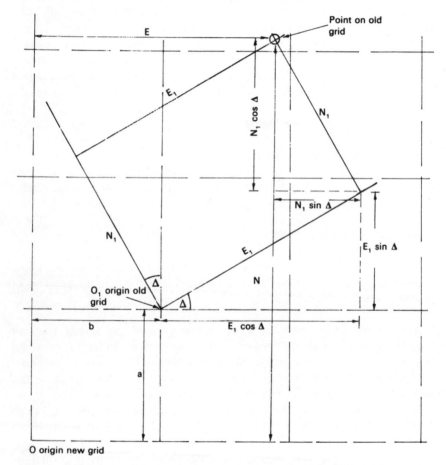

Notes   1.   The old grid shown with solid lines could be the original survey grid. The new grid shown with broken lines could be a new site grid for setting out

        2. $E = b + E_1 \cos \Delta - N_1 \sin \Delta$
             $N = a + N_1 \cos \Delta + E_1 \sin \Delta$

Fig. 3.1   Coordinate setting out − grid change calculations.

projection, in particular a Transverse Mercator Projection, for transforming measurements on the *curved* surface of the earth onto the corresponding *flat* surface of a map sheet. Using this transformation measured angles and distances have to be reduced to the projection by calculating corrections. In most cases it is normally necessary for only the distances to be corrected. For large areas consideration should be given to possible angular corrections as well. Details can be found in the pamphlet *Constants, Formulae and Methods used in the Transverse Mercator Projection*, published on behalf of the Ordnance Survey by Her Majesty's Stationery Office.

Coordinate Data

An engineer faced with a problem of fixing key points from National Grid data should take care to establish that the data supplied are correctly correlated and do not include points from different orders of triangulation with differing standards of accuracy. All data on this subject may be obtained from the Ordnance Survey, who publish two useful leaflets: OS Leaflet No. 2 *Triangulation and Minor Control Information*; OS Leaflet No. 34 *Scientific Co-ordinates for Triangulation Stations*.

The following notes, based on OS Leaflet No. 2 (by permission of the Director General, Ordnance Survey), will be of assistance in determining the problem.

About the middle of the nineteenth century a network of large triangles covering the British Isles was selected from available data and adjusted by the method of least squares for self-consistency. This network is called the Principal Triangulation. Re-triangulation was carried out between 1935 and 1962. This new triangulation was designed to provide a completely consistent network of stations of all orders. Within this framework a large number of additional control points have been established. These are known as 'minor control points'.

Density

First-order stations	40–70 km apart
Second-order stations	7–13 km apart
Third-order stations	4–7 km apart

Third-order triangulation covers the whole country (except mountain and moorland areas) and in urban areas the density is increased to give fourth-order stations 1–2 km apart.

Types of Triangulation Station Mark

(1) Triangulation pillar (familiar pillar about 1.2 m high).

(2) Surface mark – a domehead brass bolt set in rock or concrete at ground level.
(3) Buried mark – brass bolt or rod set in concrete about 0.6 m below ground level.
(4) Roof stations – domehead brass bolt.

Minor Control Points

These fall into three categories:

• permanent traverse stations
• revision points
• other minor control points.

The position of permanent traverse stations are marked on maps by the abbreviation 'ts'. They are positioned so that an instrument can be set up over them, and are marked on the ground with rivets, or rivets set in concrete, or with cross cuts on manhole frames, etc.

Data unique to each point can be obtained from the Director General, Ordnance Survey, but this does not include right of access.

The datum to be used for all levelling tasks must also be ascertained. The Ordnance Survey level datum is used for heighting on the great majority of sites in the UK.

In the UK the Ordnance Survey is responsible for the original work of determining Mean Sea Level and by levelling has produced a network of level marks (bench marks) over the whole country. A continuous process of checking and revision is carried on by the Ordnance Survey, which has its Head Office at Southampton and local offices in various parts of the country.

Mean Sea Level is more frequently referred to as Ordnance datum. Originally the datum was derived from the MSL of a dock in Liverpool, but since 1940 all levels are referred to a new datum at Newlyn in Cornwall, where the level was established from many years' recording of an automatic tide gauge.

Levels are recorded on large-scale maps in two ways, apart from normal contour lines, spot levels and bench marks. Spot levels are marked to the nearest metre and are shown by a small +. Bench marks are indicated by a small 'broad arrow' mark and are figured to the nearest 0.01 m. Up-to-date revised values for bench marks are usually available in Local Authority Offices or, if not, can be obtained direct from the Ordnance Survey. Some old maps may still be marked 'Liverpool datum'.

The letters AOD on site plans, etc., mean that level values are related to Ordnance datum (above Ordnance datum). Only very rarely (on marine or shaftworks or at Foulness Island and in some parts of the fens) will levels be below OD, in which case they will be marked with a minus sign.

Map heights should be treated with caution unless the printing date is recent. Details of revision of levels may be found in the information panel.

Bench marks are usually cut into walls and bridge abutments, sometimes gate posts. The level of the centre of the bar is the recorded height. There are other types, namely primary and rivet bench marks (see Fig. 2.7(b). Any level mark established temporarily on site is called a TBM (temporary bench mark). It may or may not be related to Ordnance datum.

Local Authorities often establish the levels of various parts of their works (e.g. manhole covers, steps, etc.) and their plans may have these values marked. Invert levels of sewers are usually referred to Ordnance datum. For obvious reasons drainage levels need this relationship.

Ordnance Survey benchmarks should be used with care. It is not unknown for walls which have benchmarks cut in them to be repaired and rebuilt with the block containing the benchmark being positioned at a different height. It is good practice on new sites to run a line of levels between a number of Ordnance Survey benchmarks in the area to check for such occurrences. Alternatively a description sheet for benchmarks in the area can be obtained from the Ordnance Survey which lists, among other things, the height of the benchmark above ground level which can then be checked. The date when the benchmark was last levelled is of importance especially in areas where mining has taken place in the past. It is not unknown for Ordnance Survey benchmarks to have dropped by metres because of subsidence. Therefore the engineer should check with the local Ordnance Survey office for the last revised value for benchmarks in the area.

Calculating with Coordinates

Calculating setting out data from coordinates or calculating coordinates from measurements is a relatively simple exercise. Some common formulae are:

$$\frac{dE}{dN} = \frac{\text{difference Easting}}{\text{difference Northing}} = \tan \text{bearing} \qquad \frac{dE}{dN} = \tan \text{bg}$$

$$\text{Distance} \times \sin \text{bearing} = dE \qquad D \sin \text{bg} = dE$$

$$\text{Distance} \times \cos \text{bearing} = dN \qquad D \cos \text{bg} = dN$$

$$\text{Distance} = \text{difference Easting} \div \sin \text{bearing} \qquad D = \frac{dE}{\sin \text{bg}}$$

$$\text{Distance} = \text{difference Northing} \div \cos \text{bearing} \qquad D = \frac{dN}{\cos \text{bg}}$$

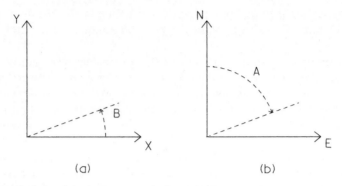

Fig. 3.2 Relationship between polar/rectangular coordinates and normal site coordinates systems.

Bearing is the clockwise angle from 0–360 from grid North except for the first formula where the resultant bearing is only for a 90° quadrant which subsequently has to be converted into a whole circle bearing. Another term often used for a whole circle bearing is *azimuth*.

A useful trick for converting between bearings and distances and co-ordinates is to use the polar/rectangular conversion found on many calculators.

Figure 3.2 shows the relationship between standard polar coordinates where the angle, B, is measured *anti-clockwise* from the x-axis and a normal survey system where the azimuth, A, is measured *clockwise* from the Northing axis. If, when using the polar/rectangular functions, the dE is entered as y and dN entered as x the resulting angle, B, will be the azimuth, A, of the line. Therefore the problem of working out which quadrant the line lies in will not arise. Similarly if the azimuth, A, is entered along with the distance, l, the resulting x should be taken as dN and the y as dE.

Control Surveys

A control survey involves the selection and marking of a number of points strategically placed around a site. Ideally these points should serve as both reference points for relating all subsequent measured detail against and also be able to be used as stations from which the setting out can be done. It may be inefficient to provide all the necessary points in a single control survey and so a primary network may be created with one or more secondary networks observed separately within it and connected to it.

There are three possible approaches to a control survey:

- Triangulation − only angles are measured.
- Trilateration − only distances are measured.
- Angles and distances measured.

Before the advent of EDM most large control surveys were carried out by triangulation with only a few carefully measured baselines providing scale. On smaller sites trilateration control surveys were often carried out using surveyor's chains, known as chain surveying. For general site detail surveys these chain surveys were often plotted graphically using beam compasses and scale rules. Now, with the widespread use of EDM, traversing is the most common form of surveying for control.

Traversing

When siting main lines intervisibility at instrument level must always be kept in mind, and at the same time the lines kept close enough to existing detail to avoid too many long offsets. Pegs should be sited where they are unlikely to be disturbed and driven well in; the centres should be marked with a nail.

When observing, it is an advantage to move round the traverse in a clockwise direction, always using the back station as a reference object. In this way, since theodolite circles are graduated clockwise, azimuths can be easily carried forward along the traverse. Traverses, like level lines, must always be closed either on a second known station or back onto the starting station. Note that the closing angle at the end station must be measured to check for angular errors.

If traverse equipment is available, the engineer can move quickly, leaving an assistant to collect and erect the stands and targets as each angle is observed. Normally, on leaving a station to move forward, the engineer should set up the rear sighting mark; in this way much time will be saved. If ranging poles are used as marks, careful plumbing exactly on line is required if angular results are to be correct. Where nail marks can be seen they should be silhouetted by something white, and a ranging pole planted to assist initial identification. In precision traversing interchangeable theodolites and targets using a constrained centring system must be used.

Traverse Computations

The first item that is required is a sketch of the traverse, drawn roughly to scale, with the abstracted observations, measured angles and reduced horizontal distances, marked on it (Fig. 3.3). If the traverse is to be computed by hand, perhaps as a quick field check, then this information is transferred to a suitable traverse computation form (Fig. 3.4). Note the staggered layout of the form to differentiate between quantities which refer to stations, i.e. angles measured and coordinates, and those which refer to traverse lines between stations such as azimuths and distances.

The procedure for using the form, using the sample survey from Fig. 3.3, is as follows:

(1) Compute the azimuth through the traverse from the starting azimuth (100−400) using the measured clockwise angles:

Azimuth = previous azimuth + measured angle − 180°

(2) Compare the azimuth computed through the traverse for the closing check, line 200−300, with the azimuth computed from the known coordinates of these two points. Note the discrepancy towards the bottom of the azimuth column.

Discrepancy = observed azimuth − azimuth by coordinates

which in this case is −10″. If this is too large then a mistake has been made. Firstly check the arithmetic on the form and on the angle and distance calculations. If this does not improve the discrepancy sufficiently than it will be necessary to reobserve one or more angles.

(3) Divide this discrepancy by the number of angles observed in the traverse, in this example 5. Each angle requires to be corrected by this value. The sign of the correction is the opposite of the sign of the discrepancy. As the azimuths have already been computed, however, it is not necessary to actually correct the angles and recompute the azimuths. Instead the azimuths can be corrected directly by correcting each azimuth by the correction factor plus the correction applied to the previous azimuth. Check that the corrected azimuth for the closing

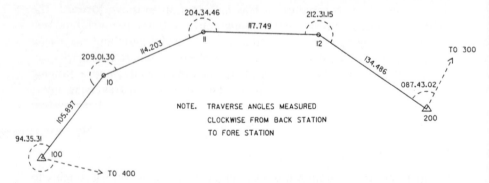

Fig. 3.3 Sketch of a sample traverse.

TRAVERSE COMPUTATION SHEET

TRAVERSE FOR New Road Survey COMPUTED BY J.R.S. CHECKED BY D.B.T. DATE 21/2/88

STN	MEASURED ANGLE	AZIMUTH	CORR	CORRECTED AZIMUTH	HORIZONTAL DISTANCE	ΔE	ΔN	EASTINGS	NORTHINGS	CORRECTIONS E	CORRECTIONS N	FINAL EASTINGS	FINAL NORTHINGS	STN
400										(mm)	(mm)			
		103.32.15	0											
100	294.35.31							959.885	1045.995	+0	-0	959.885	1045.995	100
		038.07.46	+2"	038.07.48	105.897	65.386	83.300							
10	209.01.30							1025.271	1129.295	(+4) +4	(-2) -2	1025.275	1129.293	10
		067.09.16	+4"	067.09.20	114.203	105.245	44.337							
11	204.34.46							1130.516	1173.632	(+4) +8	(-2) -4	1130.524	1173.628	11
		091.44.02	+6"	091.44.08	117.749	117.695	-3.567							
12	212.31.15							1248.211	1170.065	(+4) +12	(-2) -6	1248.223	1170.059	12
		124.15.17	+8"	124.15.25	134.486	111.154	-75.706							
200	087.43.02							1359.365	1094.359	(+5) +17	(-2) -8	1359.382 ✓	1094.351 ✓	200
		031.58.19	+10"	031.58.29 ✓										
300														
							200 should be	1359.382	1094.351					
	200→300 should be	031.58.29		Σ	472.335			dE=-0.017	dN=+0.008					
		∴ d = -10"												
		∴ +2" per angle												

Fig. 3.4 Traverse adjustment form.

line, 200–300, is now equal to the azimuth computed from coordinates. Note that no correction is applied to the opening azimuth, 100–400, as this is a fixed azimuth.

(4) Using the corrected azimuths and the horizontal distances the coordinates of the points in the traverse are computed.

(5) The computed coordinates of the closing point, 200, are compared with its known coordinates and discrepancies, dE and dN, calculated using the relationship:

discrepancy = computed coordinates − known coordinates

(6) If necessary the vector of misclosure of the traverse can be calculated from

$$\text{misclosure} = dE^2 + dN^2$$

Dividing the total horizontal length of the traverse by this misclosure will give a misclosure factor, in this example 1 : 24 860.

(7) As in (2) above if the misclosure is too large compared with the specification then something is wrong. Again all the abstractions and calculations should be checked. If this fails to improve the situation reobservation is necessary.

(8) If the misclosure is acceptable then it must be distributed through the traverse. Two corrections have to be calculated for each point, one for its Easting and one for its Northing. These corrections are proportional to the length of the traverse length to the point:

$$\text{correction in Easting} = -\frac{\text{length of traverse line}}{\text{total length of traverse}} \times dE$$

The correction for the Northing is similar.

As the coordinates of the point have already been calculated then the effect is again cumulative, e.g. in the example the correction for the Eastings of each leg, shown in brackets in Fig. 3.4, is +4 mm for the first three legs and +5 mm for the fourth leg. This accumulates to +4 mm for the Easting of 10, +8 mm for 11, +12 mm for 12 and +17 mm for 200. Again check that the adjusted final coordinates of the closing point, 200, are the same as the known coordinates of that point.

This method of distributing misclosures through a traverse is known as a *Bowditch adjustment*. It is a simple procedure to carry out with a pocket calculator but it is a non-rigorous method of adjustment as certain assumptions are made. The angular error is spread equally among all the angles in the traverse and the linear descrepancies are distributed in proportion to the lengths of the traverse lines. This may not be obvious in the above example as all the traverse legs are of approximately the same length. Wherever possible traverses should be arranged in this fashion. Unless it is absolutely unavoidable a traverse should *never* have a combination of very short lengths and very long lengths. If this becomes necessary then possibly an alternative solution to a traverse should be contemplated for the control survey.

A better method of traverse adjustment is to use the principle of least squares in which the discrepancies are distributed with respect to the *reliability* of the observations, e.g. angles with large spreads in their rounds will have large corrections applied to them. However least squares methods are rather difficult to compute on a pocket calculator but if a computer is being used it is desirable that the traverse adjustment program uses a least squares solution. There are many commercial software packages which use only a Bowditch adjustment which is a major waste of the powerful processing capacity of modern microcomputers.

If the discrepancies in (2) and (7) above are not acceptable and no errors can be found in the calculations then some quantities will have to be reobserved. There are two useful techniques which can help to pinpoints the observations which are more likely to be in error. Calculate the azimuth of the vector of the linear misclosure. If this agrees closely with, or is almost 180° different from, the azimuth of one of the traverse legs then it is possible that there is an error in the length of this leg. If no obvious agreement is found then it may be an angle that is wrong. On the traverse sketch which was prepared at the start draw a line from the closing point perpendicular to the azimuth of the vector of linear misclosure. If this line passes close to one of the traverse stations then the angle measured at that point may be in error. If neither of these techniques highlights a possible source of error then it is necessary to start reobserving the whole traverse over again.

Braced Quadrilateral

Although traversing is a very popular method of carrying out a control survey it is a geometrically weak figure. It is possible for the traverse to close within specifications but, because of errors which may compensate one another, the coordinates of intermediate stations in the traverse may have errors outside tolerance. The network can be strengthened by observing additional lines to brace the figure, say between 100 and 12 in Fig. 3.3, where possible. These additional observations will then serve as a check on some points along the traverse. However if these extra observations are to be included in the computation of the coordinates of these stations a simple adjustment procedure such as Bowditch is no longer possible. A more complicated network adjustment procedure is required.

Therefore if a control network with high reliability is required then a method other than traversing should be used.

For small, high precision control surveys a network known as a braced quadrilateral is very useful. Figure 3.5 shows the layout of a braced quadrilateral where all eight internal angles are measured. This is a very strong geometric figure as there are many angular checks, i.e. all eight angles must sum to 360° and in each individual triangle, there are four in all, the sum of the angles should be 180°. It is very simple to carry out these checks in the field and reobserve if necessary. If the closures of the triangles and the quadrilateral are satisfactory then the eight angles have to be adjusted for trigonometric consistency. There are a number of different methods that can be used to adjust the angles with the most rigorous being by least squares. A simpler non-rigorous method is to use *equal shifts*.

Equal Shifts Adjustment of a Braced Quadrilateral

The basic assumption that is made in an equal shifts adjustment is that all eight angles in the quadrilateral are likely to have random errors of approximately the same magnitude. The discrepancies in the relevant conditions are, therefore, distributed equally among the angles in question. To avoid possible confusion use of a standard form (Fig. 3.6) is recommended. The numbering of the angles (Fig. 3.5) is also very important as a mistake here will affect all subsequent work.

The first angle condition to be fulfilled is that all eight internal angles of

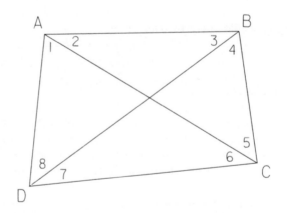

A, B, C AND D ARE STATIONS

1-8 ARE INTERNAL ANGLES

Fig. 3.5 Braced quadrilateral with angle numbering convention.

ANGLE	OBSERVED ANGLE	IST. ADJ.	2ND. ADJ.	3RD. ADJ.	PROVISIONAL ANGLES	dx	FINAL ADJUSTED ANGLES
1	019° 57' 22"	+0".75	-1".5	/	019° 57' 21".25	-3".698	019° 57' 17".552
2	066° 53' 31"	+0".75	/	+4".5	066° 53' 36".25	+3".698	066° 53' 39".948
3	055° 11' 33"	+0".75	/	+4".5	055° 11' 38".25	-3".698	055° 11' 34".552
4	016° 52' 38"	+0".75	+1".5	/	016° 52' 40".25	+3".698	016° 52' 43".948
5	041° 02' 03"	+0".75	+1".5	/	041° 02' 05".25	-3".698	041° 02' 01".552
6	066° 52' 55"	+0".75	/	-4".5	066° 52' 51".25	+3".698	066° 52' 54".948
7	055° 12' 27"	+0".75	/	-4".5	055° 12' 23".25	-3".698	055° 12' 19".552
8	037° 57' 25"	+0".75	-1".5	/	037° 57' 24".25	+3".698	037° 57' 27".948
Σ	359° 59' 54"				Σ 360° 00' 00".00 ✓		Σ 360° 00' 00".000 ✓

K1 = -6"
K2 = +6"
K3 = -18"

k = 1.00019231
dx = -3".698

Fig. 3.6 Equal shifts adjustment form.

the quadrilateral add up to 360°. Find the sum of the eight angles and then subtract 360° from this total to find the misclosure K1. The first adjustment is then −K1/8 and this is placed in the appropriate column on the form.

The second angle condition is that the sum of angles 1 and 8 should equal the sum of angles 4 and 5. Calculate the discrepancy, K2, as follows:

$$K2 = (1 + 8) - (4 + 5)$$

The second adjustment is then placed in its column with +K2/4 opposite angles 4 and 5 and −K2/4 opposite angles 1 and 8.

The third angle condition is similar but concerns angles 2, 3, 6 and 7.

$$K3 = (2 + 3) - (6 + 7)$$

+K3/4 is applied to angles 6 and 7 and −K3/4 to angles 2 and 3.

These three adjustments are then applied to the observed angles to produce provisional angles. Sum these eight provisional angles to check the preceding arithmetic by ensuring that they sum to exactly 360°.

Finally the eight angles have to be adjusted to satisfy the *side condition*. This is necessary to maintain consistency in the lengths of the sides so that if one side length is known and another side length computed from it the same answer will be obtained irrespective of which route around the figure was used in the computation. The side condition is:

$$\frac{\sin 1 \,.\, \sin 3 \,.\, \sin 5 \,.\, \sin 7}{\sin 2 \,.\, \sin 4 \,.\, \sin 6 \,.\, \sin 8} = 1$$

If the provisional angles are labelled as x_1 to x_8 then a value, k, can be calculated from

$$k = \frac{\sin x_1 \cdot \sin x_3 \cdot \sin x_5 \cdot \sin x_7}{\sin x_2 \cdot \sin x_4 \cdot \sin x_6 \cdot \sin x_8}$$

If k is equal to 1 then no further corrections are necessary. However if k is not equal to 1 a correction, dx, should be calculated from

$$dx = \frac{(1 - k)}{k \cdot \sum_{i=1}^{8} \cot x_i} \cdot 206265$$

The resultant value for dx, which is in seconds of arc, is then added algebraically to the odd numbered provisional angles and subtracted from the even numbered provisional angles. The sum of the eight final angles should be checked to be 360° and a new value for k calculated to check that it is now 1, i.e. the side condition has been satisfied.

Note that during the adjustment procedure calculations should be taken to the second or third decimal places of a second of arc to reduce the effect of rounding errors in the computation. However, when the adjustment has been completed, the final angles *must* be rounded to the nearest second (if a 1″ theodolite was used) to be consistent with the precision of the equipment used. Checks should be made that these rounded angles still satisfy the angle conditions.

Use of the Braced Quadrilateral

The eight adjusted angles will now give a very good indication of the *shape* of the figure. To obtain the *size* of the figure at least two lengths have to be measured, say one side and a diagonal. Having only one measured length would provide the scale but there would be no check on this measurement. With two lengths measured then one length can be computed from the other using the adjusted angles and checked against the measurement. Any discrepancy should be proportioned between the two sides.

The braced quadrilateral is highly recommended if setting out is required to high precision. This enables a very reliable control survey to be positioned around the area where the setting out is to be done so that the actual setting out distances will be relatively small. It is ideal when mounting bolts for prefabricated structures, such as bridge decks or steel beams, are to be set out to fine tolerances. It is also a very useful method for transferring control across obstacles such as rivers.

If larger areas are to be covered it is possible to build up a triangulation network using a series of braced quadrilaterals which have common edges (Fig. 3.7).

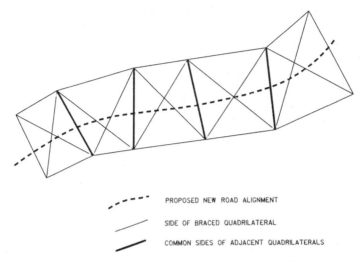

PROPOSED NEW ROAD ALIGNMENT

SIDE OF BRACED QUADRILATERAL

COMMON SIDES OF ADJACENT QUADRILATERALS

Fig. 3.7 Triangulation network of braced quadrilaterals.

Detail Surveys

Detail surveying involves recording the positions of man made and natural features which will be important for the design of the structure which is to be constructed on the site. This is normally carried out after a control framework has been established but small amounts of detail can be added to an existing site survey by relating the new detail to the already plotted detail.

Tape and Offset

The simplest method of detail surveying is by tape and offset. This method is often referred to as 'chain surveying'. Although obsolete, as surveyor's chains are no longer used, the term remains in common usage and positions along 'chainage' or reference lines are known as 'chainage' points. The term 'chainage' is also the accepted term used for describing distances along road and railway lines.

The process involves taping distances, the offsets, from either existing detail or from points along a line in the control survey. Normally the offsets are distances measured at right angles off the reference line. This can be done by using an optical square for obtaining the right angle or, if a lower precision is acceptable, by estimating the right angle by eye.

A common scale for the production of site plans is 1/500 and a survey done to this scale would have offsets to important detail restricted to not more than 10 m.

Detail lines are frequently run parallel to the faces of buildings to keep the offsets short and therefore accurate, but, when this is not done, it is customary to fix the corners by tie lines from different chainage points so that the position is not subject to the possible error of long offsets. The line of a straight fence, on the other hand, would not be the subject of such detailed measurement, except for the points at which it changed direction. While the surveyor will take accurate check measurements of buildings, he or she is unlikely to measure offsets more accurately than can be subsequently plotted; at 1/500 he or she will probably not go below 0.1 m. As this is a very low precision for taping then nylon or fibre tapes are often used for measuring the offsets as they are more easily handled and less fragile than steel tapes although a steel tape is often laid along the reference line.

A typical field sheet for detail survey is printed with a 15 mm wide column running down the centre of each page. This column represents the reference line on the ground and in it are booked the chainages at which right-angled offsets to various points of detail on either side of the line are measured. Also booked are the chainages at which a detail crosses the line, at which other lines join and from which tie lines to important detail are started. On either side of the column the surveyor sketches a plan view of the detail and notes the offsets to key points so that it can be reproduced in plan form when plotted. (See example of booking in Fig. 3.8.)

An engineer called on to carry out minor surveys should find no great difficulty. It is a mistake to try to sketch detail to scale; it is better to keep the distance between chainages in the book fairly liberal and make it more diagrammatic.

When siting traverse stations (the pegs between which the lines run), the subsequent measurement of detail should be kept in mind. For example, a line along a road should run along one edge where it can be used and offsets will be short; a parallel one can be set off the other side for detail on that side. Similarly, hedge and ditch lines often form natural guides for traverse lines.

Radial Detail Surveying (Tacheometry)

The term tacheometry means 'rapid measurement' but has come to be used to describe a method of picking up detail by measuring directions and distances, radials, out from a reference point.

This was often carried out by stadia tacheometry, a technique in which the positions of the two stadia hairs and the horizontal hair were read off a staff. When combined with horizontal and vertical circle readings distances, directions and differences in height between the instrument

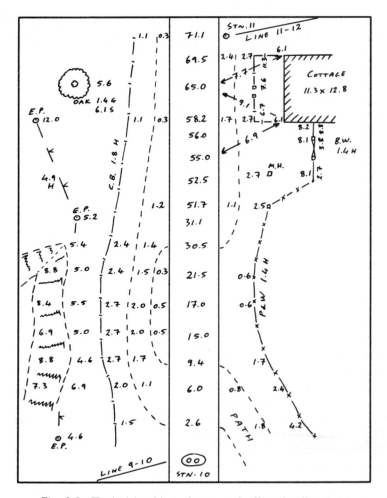

Fig. 3.8 Typical booking of tape and offset detail survey.

station and staff points could be calculated. With the advent of cheap
EDM stadia tacheometry is hardly used today. Only on very small sites
where the cost of EDM could not be justified will it still be used on
occasions.

Nowadays virtually all detail surveying using radials is carried out using
a theodolite−EDM combination, either with the EDM mounted on the
theodolite or incorporated in the theodolite to form an electronic tacheometer. If a good vantage point can be found then detail can be recorded
over a large area very rapidly and precisely. This process is even more
efficient if automatic recording of the data can be done. In this case great
care must be taken in entering the relevant coding of the feature being

measured and it is still good practice to draw a clear sketch illustrating the information which has been collected. This sketch can prove invaluable later for 'debugging' the data at the processing and plotting stage. Important points of detail should be picked up from at least two instrument stations.

As tacheometry with EDM can be carried out over long ranges, sight lengths of hundreds of metres may be used, communication between the instrument operator and prism holder can be difficult if no aids to communication are available. In this case the prism holder must be well briefed on exactly what detail is required and a system of handsignals arranged so that the instrument operator can enter the relevant code on the sketch or in the data recorder. With the ease with which much modern equipment can be used it is becoming increasingly common for the engineer to be the prism holder and the assistant the instrument operator. In this way the engineer can use his or her experience to decide what is the important detail on the site which has to be recorded.

Plotting Detail Surveys

When plotting detail from offsets it is useful to have an offset scale. This is a short scale with a true right-angled butt at the zero of the scale and, preferably, graduated at the required scale of the plot so that, when used against a long scale coincident with the reference line, the offsets measured are correctly related to the line.

For plotting radials it is necessary to have a 360° protractor large enough to be able to plot to at least five minutes of arc. It is best to plot half a dozen directions with small ticks numbered to correspond with the points and then scale off the distances to plot the detail.

Unless the levels are important it is customary to omit those which relate to points which have been picked up to denote position, otherwise the plan becomes a mass of figures. Obviously, when used for contouring, the position of each spot height thus surveyed must have its accompanying level figured neatly by its position. It is conventional to put this in the second quadrant relationship to the cross.

In many systems today the plot can also be produced automatically using data downloaded from the electronic fieldbook and subsequently processed by computer. In these cases tape and offset is not a suitable method as most of these systems are designed for computing and plotting radials.

Irrespective of how the detail was recorded or how the plot was produced the plot must *always* be taken into the field for checking and verification. This verification should not only check the metric accuracy of the plot but should also check the accuracy of the content. For example, is that tall wooden pole with wires coming from the top really a telegraph pole or is it actually an electricity pole?

Contouring

There are various ways of measuring the data from which a map or plan can be contoured, but probably the two most common are those involving direct interpolation of points sharing a common height from a large number of spot heights. In one method the spot heights are taken to a regular pattern or grid, and in the other they are taken where the skill and experience of the surveyor suggest that they will be most useful.

If recording and plotting is to be done manually the first method lends itself more easily to the measurement of the location of each spot level and subsequent plotting, while the second is quicker to do on the ground but is somewhat slower to plot, as each point must be located in terms of bearing and distance from instrument stations.

The contour interval required on many site plans is quite small, 250–500 mm, and this requires that the work must be done reasonably well if it is to be useful.

A possible task for the engineer is the gridding method, since a mesh of spot levels can be used for the direct calculation of earthwork quantities without the necessity of interpolating contours.

Level Gridding

It is often necessary to take levels on a grid over a site as an aid to exact calculations of earthwork quantities. Unless the task is approached in a systematic manner, serious errors and loss of time and money will occur.

Figure 3.9 shows one such method which will, if carried out properly, successfully prevent this. Requirements are that the grid shall be accurate and that the levels taken can be properly related to their true position on the ground.

The procedure is as follows:

(1) Select as a base line the longest possible convenient straight line between two of the original survey marks (or between two points which can be accurately identified on the plan).
(2) Set out short pegs at chosen intervals along the base line.
(3) At appropriate chainages a suitable distance apart set out lines at right angles to the base line and plant ranging poles at the grid interval on either side.
(4) The staff holder and assistant can then line themselves in on the grid lines so delineated and tape between points to position the staff at the grid intersections.

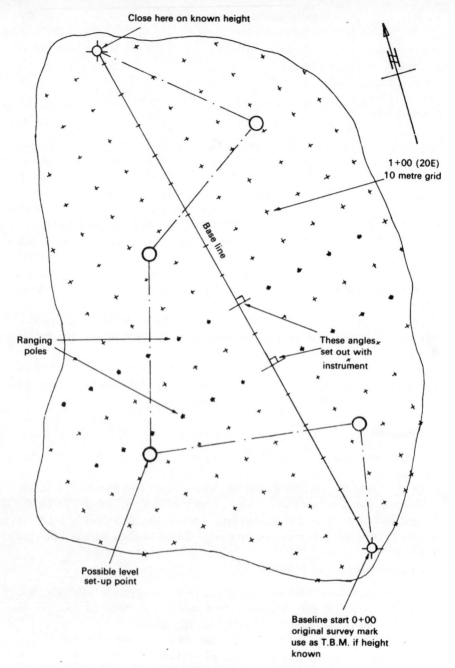

Close here on known height

1+00 (20E)
10 metre grid

Base line

Ranging
poles

These angles
set out with
instrument

Possible level
set-up point

Baseline start 0+00
original survey mark
use as T.B.M. if height
known

Fig. 3.9 Typical level grid layout.

It is absolutely essential that the plan includes a systematic method, such as that shown in the diagram, of identifying the grid intersections, and that the engineer shall carefully note in the level book the grid position of each level. Lack of forethought and carelessness in identifying

Date 22/1/80 Levels

From T.B.M. SITE OFFICE
taken for SITE LEVELS (SITE CLEARANCE)
To AREA EAST OF NEW ROAD. PATTERN C

BACK SIGHT	INTER-MEDIATE	FORE SIGHT	COLLIMATION	REDUCED LEVEL	DISTANCE	20 METRE GRID. REMARKS
1·246			121·200	119·954		T.B.M. SITE OFFICE.
1·272		1·321	121·151	119·879		C.P.
1·416		1·250	121·317	119·901		Top. Peg 8 (c.p)
	1·282		(121·32)	120·035	0 + 00	G.L. PEG 8 0+00
	1·28			120·04	0 + 00	20 L (LEFT)
	1·30			120·02	0 + 00	40 L
	1·43			119·89	0 + 00	20 R (RIGHT)
	1·72			119·60	0 + 00	40 R.
	1·46			119·86	0 + 20	20 L.
	1·51			119·81	0 + 20	40 L
	1·56			119·76	0 + 20	40 L.
	1·38			119·94	0 + 40	40 L.
	1·41			119·91	0 + 40	20 L
	1·51			119·81	0 + 40	C.P.
1·063		1·271	121·108	120·046		T.B.M. SITE OFFICE +(0·003)
		1·152		119·957		

1·063		1·271				
4·997		1·152				
4·994		4·994				
0·003						

Points to note
1. Diagram
2. Positive identification of levels
3. Check to source

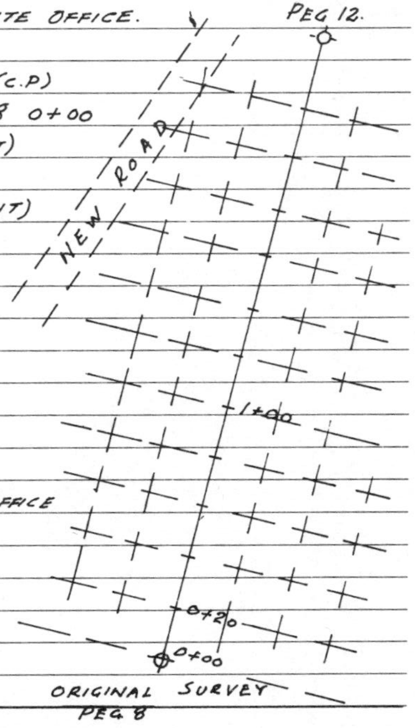

Fig. 3.10 Typical level grid fieldsheet.

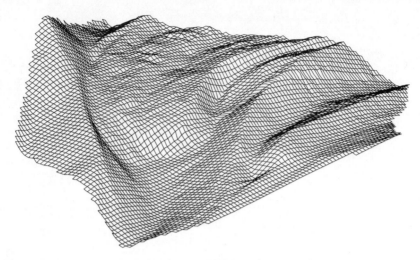

Fig. 3.11   Typical grid output from a DTM package.

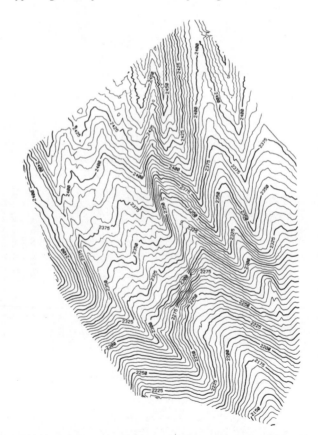

Fig. 3.12   Contour plot derived from DTM processing of grid in Fig. 3.11.

points will produce serious errors. The staff holder should be carefully instructed in the plan and if necessary given a diagram to assist him.

Level runs must open and close on known levels. On hummocky ground the staff holder should be instructed to avoid crests and hollows in order not to give false readings of the average level of the ground at the intersection. Levels must be taken right up to the boundary even if not at exact grid intersections, and their special positions noted by clear reference to the grid.

The grid size can be 10, 20 or 30 m, depending on the density of levels required. Meticulous attention to detail in booking grid levels is necessary and the notes with Fig. 3.10 should be taken thoroughly to heart.

Digital Terrain Modelling

A digital terrain model (DTM) is a description of the surface of the ground stored in a computer as a series of three-dimensionally coordinated points linked together by a mathematical expression. This model can be further processed by software to generate various types of output including contour plots and area and volume calculations. The more sophisticated modelling packages enable the mathematical surface to be altered by superimposing new surfaces, such as a new road alignment with embankments, and cuttings, and can generate perspective views, engineering drawings and setting out data. Figure 3.11 illustrates an isometric view of a typical DTM and Fig. 3.12 shows a contour plot derived from it.

There are numerous DTM packages available for civil engineering applications but there are only two main approaches to the creation of the computer model:

(1) using a regular grid of height points, or
(2) using an irregular disposition of three dimensional points from which triangles are generated.

Grid based DTM packages reproduce mathematically the process carried out by an engineer handling the results of a grid levelling survey. Therefore grid levelling results make an excellent input for such a package. The main disadvantage of this approach, as in grid levelling itself, is that a regular grid may miss significant changes in the surface of the ground unless the grid points are very close together in which case vast amounts of data have to be collected and processed.

Random point based DTM packages are more suited to data collected by radial methods of detail surveying and produce a better representation of the surface as the data can be collected at all significant changes in the surface. Some grid based packages can accept random data as input but then interpolate a grid from this information thus smoothing out the surface somewhat.

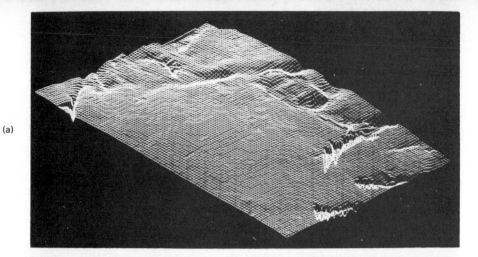

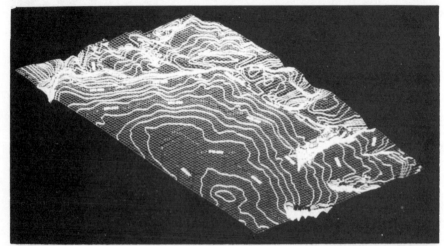

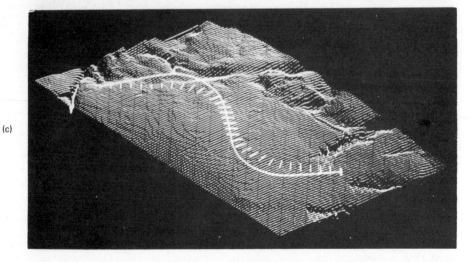

Fig. 3.13 (a) A typical grid display of a digital terrain model.
(b) The same grid display with derived contours superimposed.
(c) The same grid display with planned road alignment with cross sections superimposed.

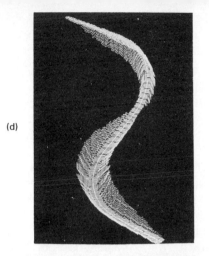

(d)

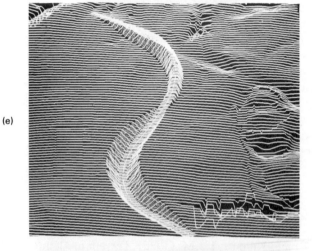

(e)

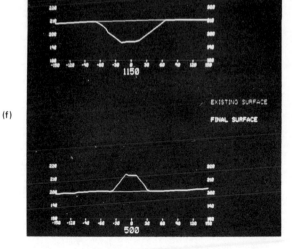

(f)

Fig. 3.13 (d) Detail of planned road with cutting and embankment.
contd. (e) New road detail superimposed on existing terrain.
(f) Typical derived cross sections of planned new road.
(All photographs courtesy of Intergraph (UK) Ltd)

On large projects, such as major highway schemes, DTMs are often generated using photogrammetric methods usually using aerial photographs of the area. This involves producing three dimensional information from measurements off pairs of stereo photographs of the area concerned.

In whatever way the input data are generated the engineer is usually particularly interested in the various types of data which can be output from such packages. The simplest packages, which can be run on small desk top computers, will only generate contour plots and simple area and volume calculations. More sophisticated DTM packages are tailored for specific engineering design tasks. A very common task is in route planning and highway design. Figure 3.13 illustrates various types of output which can be generated with a typical system. Output types not shown are detailed design drawings and setting out data. It is possible to select stations from which setting out will be done and the program will then generate all the necessary data for the setting out of the key points of the design. If no specific instrument stations are selected then the three dimensional coordinates of all the key points can be output for later use either in another software package or downloaded directly into the memory unit of certain electronic tacheometers, electronic fieldbooks or portable computers.

Petrie and Kennie[1] have produced a useful survey of various DTM packages for civil engineering with their advantages, disadvantages and applications.

Reference

(1) Petrie, G., and Kennie, T.J.M. (1986) 'Terrain modelling in surveying and civil engineering'. *Proceedings of a Conference on 'State of the Art in Stereo and Terrain Modelling'*, The British Computer Society, 28th May 1986.

Part Two
Setting Out

4 The Process

Setting out means the provision of marks, levels, profiles and other information which will enable other members of the construction team to carry out their work so that the result is a true interpretation of the contract documents in terms of position, size, shape and level.

It may vary from the simple provision of corner pegs for a small bungalow to all the marks and control information for a motorway, an industrial complex or a new town.

It involves marks and control for any form of new construction, including excavation, filling, tunnelling, building, bridging and marine works.

Broadly, it divides into three phases:

- Initial phase
- Intermediate or construction phase
- Finishing or final phase.

It is a continuous process which starts sometimes before a contract is let and needs constant updating of information, reprovision of marks, levels, etc., and final checks.

There are broadly three types of works:

- Highways (including railways)
- Bridges and buildings
- Drainage works.

Sources of Information

(1) *The Site Plan* will show the general arrangement of the new works in relation to the present detail and form of the site. Site plans may vary in scale quite widely but the most common is probably 1/500.
(2) *Setting Out Drawings* show the site arrangement in less detail but include key dimensions for setting out centre lines, main buildings, etc.
(3) *Detail Drawings* show the detail of various parts of the work to a larger scale and include dimensions.

(4) *The Specification* may include written detail of certain aspects of the construction which will affect measurement not shown on plan, e.g. drainage bedding and haunching.

(5) *Computer Print-outs* show a variety of data such as highway curve, level values, setting out data, etc.

(6) *Schedules* similar to the above show collated values for such items as manhole invert levels, centre line and margin levels, etc.

(7) *Tables* provide curve or super-elevation data for highways or railways.

(8) *Survey Data* relate to the original site survey or to the National Grid in terms of coordinates or Mean Sea Level (Ordnance datum).

Throughout the following chapters certain conventions in common use will be included in suggested methods.

All or nearly all methods depend on skill in angular and linear measurement for their successful execution and only to a limited extent on a more than ordinary ability in mathematics. Ability to read and understand drawings is of paramount importance, as is the cultivation of order and method in setting out activities.

All methods should include independent checks and the clearest possible information for others. In this latter respect the engineer should find out the ways in which other people will work from the information provided.

5 Marks and Profiles

There is a reasonably well established code of custom and practice relating to marking of pegs, profiles, etc., which is here described in detail, under three headings:

(1) Roadworks
(2) Drainage works
(3) Building works.

Information conveyed must be sufficient for the phase of work, and the process must guard against the destruction of the marks as the work proceeds by the use of such arrangements as reference pegs, guard posts, offset pegs, etc. Too early provision in other than the initial phase may be as time-wasting as too late. Conventions which have arisen are based on sound practical principles and are, therefore, not to be ignored if unnecessary extra work is to be avoided and production is to be maintained.

No universal colour code for peg marking exists but there are sufficient combinations of basic colours available for practical site coding to be established and known to all concerned on any one site. Colour conventions which have been used are:

Yellow	master pegs
Red	foul drainage
Blue	surface water
White	buildings
Black and yellow	guard posts

The quality of marking pencils is now so good that all important information can be given by proper marking of plain or white pegs or profiles.

Road Works

Initial Phase

(1) Centre line pegs at chainage intervals (every 10 m). The convention for marking is that whole hundreds (i.e. chains) are shown first and small divisions are indicated after a plus sign, e.g. a centre line peg

370 m from the start will be marked 3 + 70. It is also a convention that these marks are on the side facing the start. Since they are set with care they may carry a nail or cross mark indicating the exact line.

(2) Intersection point pegs, i.e. pegs marking points of intersection of straight lines of road, are marked with identity, curve number and intersection angle.

(3) Tangent point pegs marking points where line of road changes from straight to curve are marked with chainage, TP (or TC) and curve number and radius. A variation of this occurs where the main curve has a transition lead-in, in which case it will be marked TS (tangent to spiral), the chainage, with, ideally, the length of the transition, e.g. LS 200 m (Fig. 5.1). A consequent variation will be at the point of change from transition to circular curve where the peg will be marked SC (spiral to curve), with chainage, radius and number of circular curve (Fig. 5.1).

All of these pegs should be guarded initially, and immediately referenced to avoid the protracted and time-wasting process of having to reset completely if disturbed. Reference pegs must be set well outside the area of proposed works and so arranged that the marked peg position can be relocated to its original accuracy. Reference pegs should themselves have guard posts (Fig. 5.2).

There are occasions when key pegs on road alignments are set out many weeks or months ahead of construction, and in this case there are two further conventions to note in relation to tangent point and intersection point pegs.

(1) Intersection points are marked additionally with two sloping pegs in the form of an inverted V (Fig. 5.3).

(2) Tangent points are marked with two additional pegs, sloping the other way in the form of a V (Fig. 5.3).

Their purpose is thus clearly defined long after any markings may have weathered away.

Side Width Slope Pegs (Batter Pegs)

Since any new road formation is certain to involve cut and fill, there is an urgent need to define the limits of both before construction starts. A useful convention is that side width pegs are set at every cross section (e.g. 10 m) a standard distance outside the line of cut or inside the line of fill.

The pegs are marked with chainage and rough depth of cut or fill. It is an advantage to use long pegs on fill sections so that initial dumping is well confined and the pegs are not easily covered. In addition they serve to become part of the later stage of erecting batter rules.

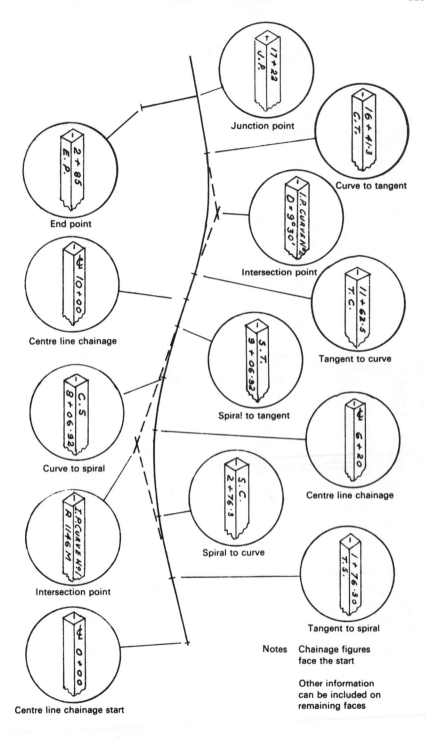

Fig. 5.1 Typical road centre line pegs.

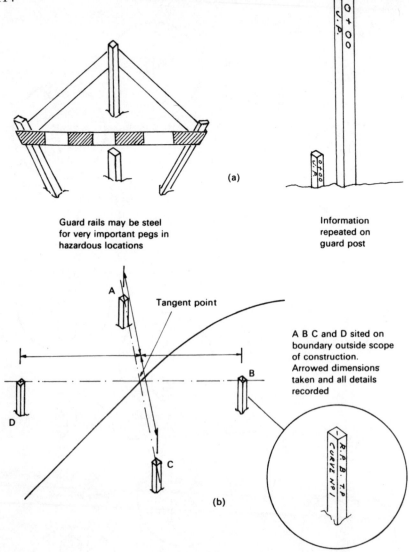

(a)

Guard rails may be steel
for very important pegs in
hazardous locations

Information
repeated on
guard post

Tangent point

A B C and D sited on
boundary outside scope
of construction.
Arrowed dimensions
taken and all details
recorded

(b)

Fig. 5.2 (a) Guard posts.
(b) Typical arrangement of reference pegs for key centre line pegs.

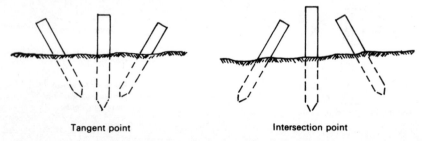

Tangent point

Intersection point

Fig. 5.3 Long-term key point marks. The significance is clear long after markings
have weathered.

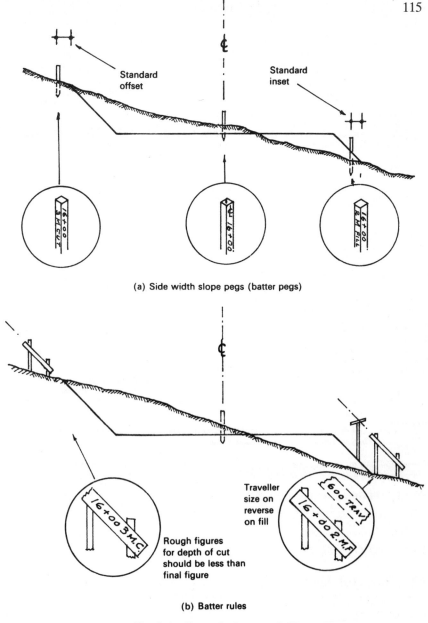

(a) Side width slope pegs (batter pegs)

(b) Batter rules

Fig. 5.4 Control of cut and fill.

In fill sections profiles are erected on the centre line with a Tee piece showing the extent of the fill (it usually includes the surcharge, i.e. the amount allowed for settlement). These profiles are subject to damage during spoil dumping but should be maintained to control not only the depth but also the main line, which can easily become offset (Figs. 5.4 and 5.5).

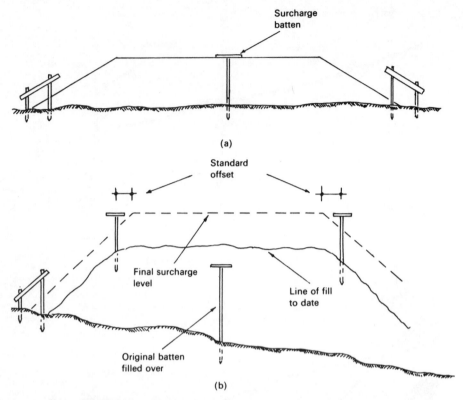

Fig. 5.5 Typical arrangements for control of fill. (a) Surcharge batten on centre line for light fill. (b) Progressive control on sections of heavy fill.

Construction Phase

The line of cut and fill having been marked by initial phase pegs, there now arises the further requirement of final bank trimming and sloping of the formation.

Batter Rules

Banks are controlled by batter rules which are set with the side width slope pegs to show and control the slope of the bank (Fig. 5.4). They are marked with chainage and rough depth of cut and fill. As a precaution against damage or differences in different parts of the contract, they can also be marked with the batter, e.g. 2 to 1.

Formation Profiles (Fig. 5.6)

These are set back a standard distance from the finished kerb line at every cross section and carry a crosshead for use with a traveller to control the level of the alignment. The level of the crosshead relates to finished road level. They are marked with chainage and distance down to FRL. They should be set square to the alignment and the offset measured to the nearest face.

Crossfall Profiles (Fig. 5.7)

These are required in the finishing stages of forming the carriageway and are erected at every cross section to control the various crossfalls which may be required. If the verge has a different fall from that of the carriageway, each profile will have two heads of different colours for use with

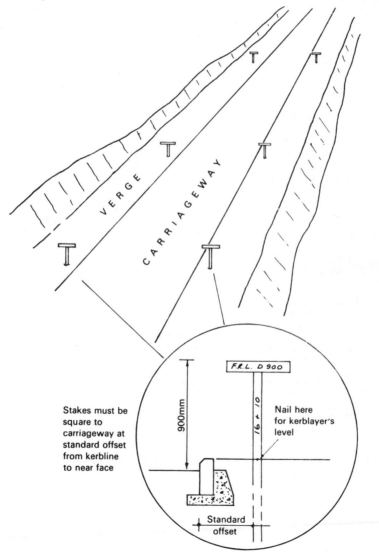

Fig. 5.6 Roadworks level control − formation profiles.

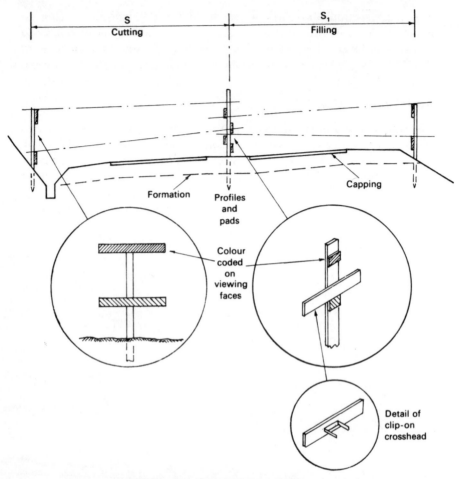

Notes   1. S and S₁ standard dimensions on cut and fill
        2. Pads and profiles colour coded for carriageway and hard shoulder
        3. Separate travellers for cut and fill sections and slopes
        4. Clip-on crosshead used only when boning
        5. Special marks to pads where construction changes
        6. When setting out prepare a schedule of levels first
        7. Set stakes first then level marks
        8. Fix crossheads and pads by direct measure from level marks
        9. Mark travellers and colour code
       10. Disseminate information

Fig. 5.7   Roadworks − profiles and padstakes − cross fall control.

different travellers. They face across the carriageway and, in addition to being colour coded, are marked with:

- Chainage
- Crossfall
- Traveller length.

Padstakes (Fig. 5.7)

On motorway construction with dual carriageways there will be at least four crossfalls per section; the cross-boning is done separately for each carriageway, but uses only one stake on the centre reservation. As this would have a number of crossheads it is more convenient to substitute short wooden pads which control the level of clip-on crossheads when required. They are colour coded to match the profiles and may have up to six different colours. The pads are fixed on the side of the stake nearest the profile to which they relate. They are marked with chainage. All crossheads are marked on the back with details of traveller length.

Additional Pegs

On estate development or industrial complexes which involve a network of minor roads there are additional centre line pegs which mark intersections and end points. They are marked with the following:

Junction points	JP and number
End points	EP and number

They may be additionally marked with chainage on the appropriate faces. (See Fig. 5.1.)

Drainage Works

Initial Marks

Drainage systems (other than land drainage) of small and medium diameter normally run point to point in straight lines between manholes not more than 100 m apart. It is normal practice, therefore, to locate points of change of direction and gradient by manhole marking pegs (Fig. 5.8). They are marked with:

- The manhole number
- The diameter
- Type of system.

Since excavation work will immediately displace them, they are referenced by offset pegs outside the scope of excavation and spoil dumping.

Reference pegs are marked with manhole number, diameter, type of system and offset distance. They are normally in pairs at right angles to the drainage line (Fig. 5.8).

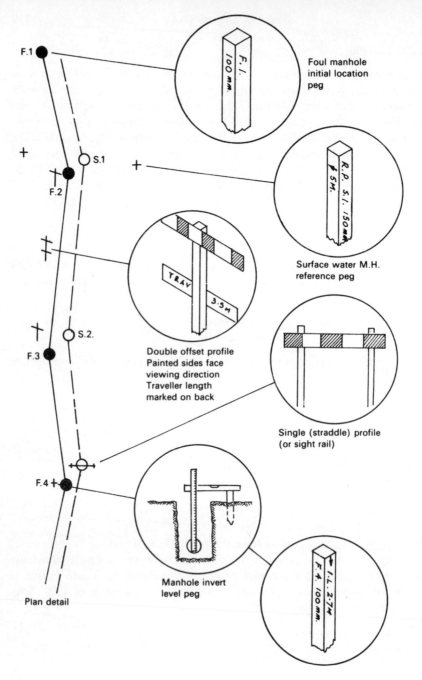

Foul manhole
initial location
peg

Surface water M.H.
reference peg

Double offset profile
Painted sides face
viewing direction
Traveller length
marked on back

Single (straddle) profile
(or sight rail)

Manhole invert
level peg

Plan detail

Fig. 5.8 Typical peg and profile marks (drainage).

Construction Marks

These take the form of profiles for use with travellers to control the depth
and gradient of the trenches. They may be either single stakes with cross-

heads at predetermined level, offset from the trench to allow the machine to work, or double stakes astride the line supporting a sight rail (Fig. 5.8).

The latter kind may additionally be set back along the line to provide room for a machine to work. In either case they are marked with the length of traveller to be used. This is understood to refer to the invert level of the drain. The travellers are lengthened for additional excavation for bedding. The marks can be colour coded in black, blue or red with white for ease of identification and boning. The coloured side should face the direction from which it is viewed and the traveller length should be marked on the plain face.

Where convenience and changes of grade dictate they may be double headed, in which case the sight rails are coloured to correspond with the distant sight rails so that confusion does not occur in use (Fig. 5.8).

Sometimes after digging is complete small reference pegs are placed near the manhole excavation to control invert levels. They should be near enough to be used with a straight edge and spirit level for transfer of level to the trench bottom by direct measure.

They are marked with the manhole number, diameter and a level mark and dimension down to invert level (Fig. 5.8).

Building Works

Initial Marks

These are concerned with the shape and position of buildings in terms of the external corners. They are, therefore, generally self-explanatory and need no special markings as they are rapidly superseded by construction marks or profiles outside the limits of the groundworks necessary.

An exception to this may be a level peg in the vicinity of each building associated with finished or structural floor level (Fig. 5.9).

Construction Marks

There are broadly only two kinds of construction mark: profile boards and offset pegs. Of these, profile boards are markedly superior, in that level control of work can be exercised as well as position.

It is conventional to set back profile boards on the extensions of the lines of the finished building in such a way that the intersection of stringlines between marks on them denotes the corners of the building or associated foundation works.

In addition, they are set at a fixed relation to structural floor level (SFL) so that level control can be exercised across the site of the building by boning. The traveller length for SFL is marked on the face. Saw cuts

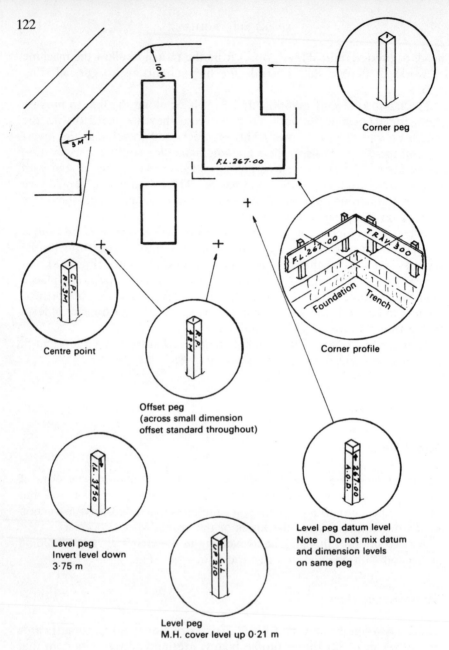

10 M

3 M

F.L.267·00

Corner peg

C.P.
R=3M

Centre point

R.P.
1·8 M

Offset peg
(across small dimension
offset standard throughout)

F.L.267·00   TRAV.300

Foundation   Trench

Corner profile

I.L. 3·750

Level peg
Invert level down
3·75 m

267·00
A.O.D.

Level peg datum level
Note   Do not mix datum
and dimension levels
on same peg

C.L.
UP 210

Level peg
M.H. cover level up 0·21 m

Fig. 5.9   Typical peg and profile marks (building and drainage).

are made to denote the extension of the foundation lines, etc. (Fig. 5.10).

Profiles should be set back sufficiently to be clear of excavation and spoil.

Offset pegs are an inferior substitute, which, as their name implies, are offset from the corners of a building, usually across its smaller dimensions (Fig. 5.11).

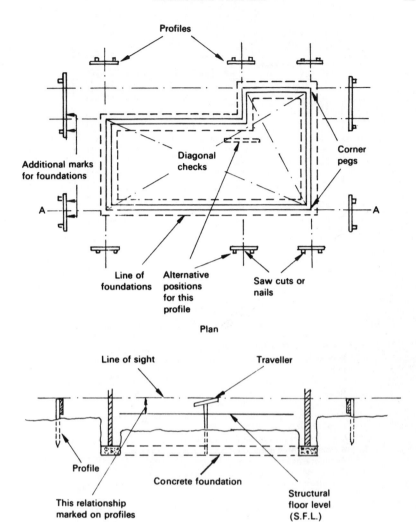

Profiles

Additional marks
for foundations

Diagonal
checks

Corner
pegs

A ——— A

Line of   Alternative
foundations   positions   Saw cuts or
               for this   nails
               profile

Plan

Line of sight   Traveller

Profile

This relationship
marked on profiles

Concrete foundation

Structural
floor level
(S.F.L.)

Section AA

Fig. 5.10   Building marks and profiles.

Additional Marks

Examples of these are level marks and centre points. It is often necessary on construction sites to have local reference to level datum in convenient areas. They may be TBMs (temporary bench marks) for general use or a datum level for nearby construction. The level referred to may be the top of the peg or bolt, or a scribed or marked line on its face or some convenient surface.

If the information is concerned with level only, this is indicated by the marking being in terms of MSL, e.g. 21.875 AOD (Fig. 5.9). If, however, the level mark bears a fixed relation to some construction level, it should

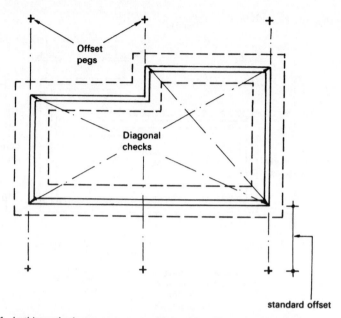

standard offset

Notes 1. In this method corner pegs are referenced by offset pegs across the
          least dimensions
       2. An offset peg is more prone to disturbance than a profile frame which
          is very stiff in one direction. Pegs should therefore be firmly driven and
          protected
       3. Care must be taken in the exact setting of pegs in both horizontal planes
          since the size of the building depends on this, unlike profiles which define
          the line of the building in only one plane
       4. Separate arrangements for level control by pegs or engineer levels given
          from time to time
       5. Careful checks on operatives' measurement must be done as the size of the
          building is more dependent on their work in this method
       6. Sequence as before including diagonal checks
       7. The overall dimensions of the reference pegs must be checked
       8. The offset should be marked and should be standard throughout

Fig. 5.11 Setting out buildings – Method B.

be marked in dimensional terms, e.g. Down (or Up) 3.475 m (Fig. 5.9). It
is better not to combine the two types of marking on one peg unless no
possibility of error occurs.

In small road works, approaches, etc., many small radius curves are
involved which can often conveniently be set out by scribing with a tape
from a centre point. They should be marked CP with radius of curve (Fig.
5.9). This device is not useful over radii in excess of 30 m, owing to
scribing difficulties in the absence of tapes over this length.

General Hints

The whole purpose of marks, pegs, profiles, etc., is to provide information
from which others can work with a maximum of certainty and a minimum

of inconvenience. It is worthwhile, therefore, to take some care to make them conform to the following rules. They should be:

(1) legible
(2) unmistakable in meaning
(3) weatherproof
(4) secure against movement or damage.

Marks on wood are enormously improved if the surface is smoothed beforehand. When marking pegs, always start at the top and write downwards. Use a weatherproof marker or crayon. Use contrasting colours where possible, e.g. black, blue or red on white; white or yellow on black.

Be systematic, e.g. adhere to standard markings, offset distances, etc. When this is not possible, take pains to bring the difference to notice. Keep records of setting out works and details of marks provided, and make sure that all concerned understand the markings and conventions used. Erect posts to guard against damage before usefulness is over, or where identification is difficult.

Keep marks up to date and checked from time to time and remove when no longer valid.

Special Marks

On multi-storey work with the implied deep foundations it may be necessary to reference the corners with permanent or semi-permanent marks set in concrete as a frame of reference for future marks on a surrounding profile. Such marks should be guarded against damage by plant and machinery since this type of construction makes early demands for accuracy in setting out at foundation level, e.g. lift shafts, etc. (Fig. 5.12). In this type of construction, as in steel-framed buildings, there is also a demand for accurate marks on the structural floor to control column spacing and verticality.

These marks are best made by setting small steel plates (100 mm square) in the green floor slab and subsequently marking them with a centre punch or a small drilled hole for the accurate marking of line or position. They denote the intersection of grid lines which relate directly to column spacing. The intersections may not, however, indicate column centres but offset positions to facilitate erection and checking of formwork. It is more usual for them to be offset (Fig. 5.12).

In multi-storey work the positions are pre-planned to coincide with purpose-made apertures in successive floors to permit control of verticality by optical upward plumbing methods. Position in all cases should be associated with a figure so that only code marks or numbers are necessary.

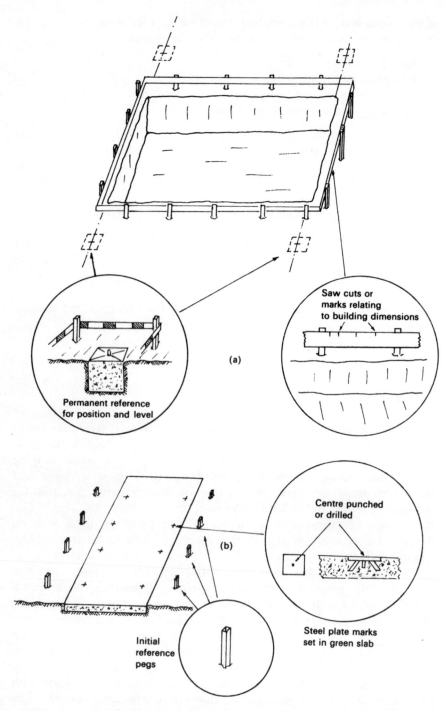

Fig. 5.12 (a) Typical marks for deep foundations and high rise building.
(b) Typical marks for column bases or vertical control.

6 Setting Out Road Centre Lines

The task of setting out road centre lines may vary between very wide limits, from some kilometres of motorway to quite short stretches of new road in a new construction site. The engineer must decide which of the methods suggested here is most suited to the job he has to do. The conventions about marks and information will, however, always apply.

Preparatory Work

(1) Examine the site plan and setting out drawings for information on the location of key points:

- The start point (0 + 00)
- Intersection points of straights
- Tangent points to curves.

(2) Look at the specification for information about accuracy of location.
(3) Make dimensioned sketches in a field book of the location of original survey marks near to key points. If there are none near, sketch in the permanent detail.
(4) For motorways which may require location in terms of the National Grid, look for local trigonometrical control. The local authority may already have caused points to be located near to the final line in terms of the Grid and details will exist of any such marks.
(5) Check curve data (site plan, print-out or schedule).
(6) Make a simple plan of reconnaissance taking construction priority into account.
(7) Check availability of instruments and prepare list of stores required.

Walking the Course

(1) Take some equipment with you, e.g. ranging poles, pegs, tapes and marking pencils.

(2) Locate key points (IPs at this stage) in relation to detail. Plant ranging poles and drive rough location pegs.
(3) Walk the straights, taking note of visibility between points and measurement difficulties, e.g. slopes, obstacles.
(4) Note requirements for immediate clearance of trees, hedges or undergrowth on the centre line.
(5) Identify points for trigonometrical control and mark.
(6) Make detailed plan for exact location of key points.

Planning Factors

(1) Priority of work
(2) Method of fixing key points
(3) Data and resources available
(4) Materials and instruments required
(5) Level control
(6) Measurement of IP angles
(7) Calculation of curve data
(8) Curve ranging
(9) Reference pegs required
(10) Line of sight clearance.

Fixing Key Points

There are two possible methods, of which the first is more likely to be used:

(1) Fixing by angular and/or linear measure from permanent detail.
(2) Fixing in terms of coordinates.

Fixing from Nearby Permanent Detail

The setting out drawing should show details of location in relation to permanent detail or original survey marks. A sketch or tracing should be made of the location of IPs, including any dimensions shown. In default of any such information, the dimensions must be scaled from the plan with sufficient extra information to relate any distortion to measured distances on the ground.

The aim should be to locate the point by either angular or linear measure, or both, so that its position is 'locked', i.e. check that dimensions to at least three points of permanent detail all agree within permissible limits (Fig. 6.1).

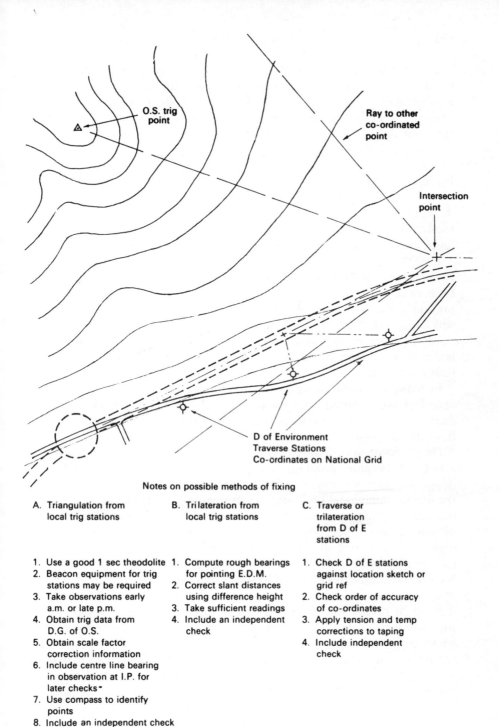

Notes on possible methods of fixing

A. Triangulation from local trig stations	B. Trilateration from local trig stations	C. Traverse or trilateration from D of E stations
1. Use a good 1 sec theodolite	1. Compute rough bearings for pointing E.D.M.	1. Check D of E stations against location sketch or grid ref
2. Beacon equipment for trig stations may be required	2. Correct slant distances using difference height	2. Check order of accuracy of co-ordinates
3. Take observations early a.m. or late p.m.	3. Take sufficient readings	3. Apply tension and temp corrections to taping
4. Obtain trig data from D.G. of O.S.	4. Include an independent check	4. Include independent check
5. Obtain scale factor correction information		
6. Include centre line bearing in observation at I.P. for later checks ·		
7. Use compass to identify points		
8. Include an independent check		

Keep clear records

Fig. 6.1 Roadworks – fixing motorway keypoints on National Grid.

Peg and guard post should be driven in, the position of the centre checked and any discrepancies between site plan measure and ground measurements recorded. Since key points will in the first instance be IPs, chainage marks will not be involved. It may be both convenient and time-saving at this point to put in reference pegs.

Fixing from Coordinates

A check should be made with map and data from Ordnance Survey or local authority on local trigonometrical control. This may be some distance away and a plan for tying-in will have to be made.

If EDM is available trilateration may be done rather more quickly and easily than triangulation. In certain types of terrain, closed traversing from control to key point may be necessary. Visibility problems may prevent a direct fix of a key point by triangulation or trilateration, in which case an arbitrary position nearby which overcomes this can be fixed. Transfer to the final point can be done by normal angular and linear measure after calculation of the difference between arbitrary and final coordinates.

In fixing by triangulation or trilateration, three points of known co-ordinates are required, two for the fix and one for the check. Careful plumbing of a beacon or mark over the key point is essential for observation from distant points.

On all major highway projects the setting out data is now produced by a suitable computer program. The provisional output from the highway design program is normally a list of coordinates of all the key points in the design. Additional programs or subprograms are available to produce the actual setting out data for a particular situation. These will probably include, among others, options for setting out by bearing and distance from certain control stations, theodolite intersection from two or more control stations or setting out from chainage points on the road alignment itself. One of the advantages of having a small computer or certain types of electronic tacheometer on site would be the ability to rapidly alter the method of setting out if, say, a line of sight had been blocked. In a situation such as this, new setting out data for different control stations could be quickly generated.

However on a number of small projects these computing aids may not be available and so the sections that follow detail a number of ways in which setting out data can be calculated on site using pocket calculators and standard tables. It must be emphasised that these methods are only approximate when calculating and setting out transition curves because of certain assumptions that are made in interpolating values from the tables. If higher precision is required then the relevant formulae for transition curves can be used to obtain better data. These formulae are listed in *Highway Transition Curves (Metric)* produced by The County Surveyors' Society. With a reasonable calculator, especially a programmable one, this should not prove difficult.

Chainage Pegs (Straights)

The plan position of the chainage marks will, at this stage, be accurate only in terms of the plottable accuracy at the scale of the drawing, e.g. probably not better than the nearest 150–300 mm at 1/500 scale. The actual positions will be more accurate than this as they will be measured from the start (0 + 00) by normal linear measurement producing an expected accuracy of at least 1/10 000.

Two courses are open to the engineer controlling the setting out: the centre line can be measured from (0 + 00) and carried right through the job, or it can be started at the most convenient point or points and any discrepancies between surface measurement and plotted chainage can be absorbed in the various sections of the job.

If the road begins with a curve, this must be calculated and the tangent points set out to establish the chainage for the first tangent point on the first straight. If the ends of the first straight are intervisible, the procedure is as follows:

(1) Set up a theodolite over the starting point chosen (of known chainage) and lay on the far mark erected over the further key point on the straight. For distant shots, a carefully plumbed ranging pole will do, but if traverse targets are available these are to be preferred. On close shots it will often be possible to sight on the nail mark on the far peg. With a telescope of 28 to 30 times magnification, a nail silhouetted against a light background (a postcard) can be picked up at 500 m. A ranging pole should be planted to assist in initial identification.

(2) Using a steel tape (30 or 100 m), set in a peg at the first multiple of 100 m chainage (e.g. 3 + 00). Set the next pegs at chainages suitable to the tape in use:

30 m tape   3 + 30, 3 + 60, 3 + 90, etc.
100 m tape   4 + 00.

The practice of hooking the loop of the tape over the nail mark is not suitable for this work; two experienced assistants are required if speed and accuracy are to be obtained.

(3) Line in for assistants with a ranging pole a short distance beyond the end of the tape.

(4) Assistants measure exact distance (using correct tension) and a peg is driven exactly on line.

(5) Assistants check that top of peg is at correct distance and scribe a mark at this distance across the top.

(6) A sighting mark is now held on this scribed line (a yellow pencil makes an excellent mark) and is moved exactly into line by directions from the engineer at the instrument. Clear signals should be given. A nail is now driven into the peg at the final position, leaving about 5 mm protruding.

(7) This is now checked for distance by the assistant and for line by the instrument operator. Adjustments can be made by firming the ground round the peg with a sledge-hammer rather than by pulling the peg.

(8) If chainage pegs are required at every 10 m these can be put in by a slightly less exacting process between main pegs later.

If an EDM of sufficient precision is available then this can be used for measuring the 100 m chainages. As it is difficult to hold the prism exactly vertical a good technique is to position two pegs on line, one slightly less than 100 m from the instrument and one slightly more than 100 m from the instrument. Set a tripod up over each in turn and measure the distance to each peg. The 100 m chainage point can then be positioned by laying a tape between the two pegs and positioning a third peg at the correct location. This third peg should then be checked from the instrument station by setting a tripod and prism over it. However it will probably prove quicker to use a 100 m steel tape unless the ground is very uneven. For the intermediate pegs, 30 m and 10 m, taping will be much more efficient than using EDM.

If the ends of the straight are not intervisible, and cannot be made so because of the shape of the ground, an initial line-finding process is necessary.

(1) If the setting out drawing gives an angular relationship of the straight with some reference mark, then set and turn off this angle on the theodolite. Sight in a mark on the straight as far distant as possible. If the far end of the straight is not visible from this intermediate mark, the process will have to be repeated there, and a further intermediate mark set up.

(2) Set up theodolite at the intermediate mark, lay back on start peg and transit the telescope on both R and L faces to sight in mark at the far end of the straight. In the unlikely event of this mark being exactly on line, chainage pegs can be set out as described above.

(3) If, as is almost certain, this line produces an error at the far end of the straight, the assistant should be instructed to measure the error. The intermediate peg can now be moved to a new trial position by taking a proportion of the line error in relation to the approximate distance between the stations.

(4) A quick check with the instrument at this new position will make the last minor adjustment easy. It should be done as follows:

(i) Lay on start, measure angle to far end; this will be a small amount + or − of 180°. If the intermediate mark is approximately half-way between the ends, this small angle will be shared equally between the angles at both ends, i.e. it will be twice the line error at either end. If the distances are not equal, the line errors will be in inverse proportion to the distances as parts of the small angle.

(ii) From the example shown in Fig. 6.2, the angle measured at C_1 will be a. Angle x+y will be 180°−a or 180°+a depending on whether the trial point is to the right or left of the correct line AB. x and y are in inverse proportion to the distances AC and BC.

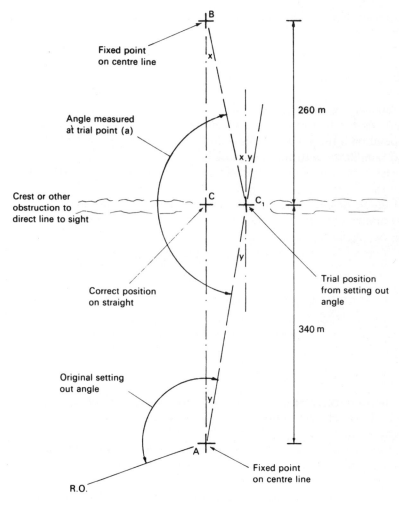

Fig. 6.2 Roadworks − lining in on straights.

$$y = \frac{CB}{AB} \cdot (x + y)$$

The correction C_1C is

$$C_1C = AC \cdot \tan y$$

Another peg, or nail if C_1C is small, is positioned, the instrument moved to C and the angle ACB checked for 180°.

Numerical example (from Fig. 6.2):

If $(x + y)$ $= 23''$

     y $= \dfrac{260}{600} \cdot 23 = 10''$

Therefore $C_1C = 340 \cdot \tan 10'' = 16mm.$

Curve Ranging

In highway work instrumental methods are used, but there are other methods for less exacting tasks, using simple linear measurement. All depend on a thorough understanding of the simple geometry of the circle and familiarity with the terms used.

There are two methods of denoting curvature: the normal radius method and the degree of curvature method.

Degree of curvature means that a 5° curve is that curve of which a 100 m arc subtends 5° at the centre of the curve. The radius of curvature can be calculated from

$$r = \frac{100 \times (360 \div D)}{2\pi}$$

where D is the degree of curvature. Therefore a curve with a degree of curvature, D, of 5° has a radius of 1145.916 m (Fig. 6.3).

Degree curves are very convenient for calculation of setting out data since the deflection angle from the tangent for a 100 m chord is equal to $\dfrac{D°}{2}$, i.e. for a 5° curve, 2°30′.

The County Surveyors' Society, in *Highway Transition Curves (Metric)* (The Carriers Publishing Co. Ltd), give a list of preferred radii for highway work in degree terms. Transition curves are a feature of all motorway work and in the UK are likely to be based on this publication. They are also increasingly used on low speed roads to make the passage of traffic easier.

In general they are used in conjunction with a normal circular curve, and form a gentle lead-in from the straights (tangents) to the desired

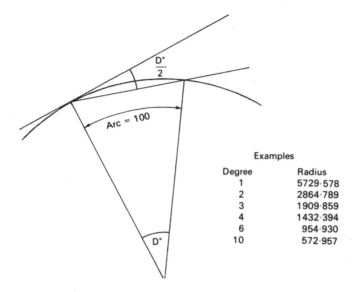

Examples

Degree	Radius
1	5729·578
2	2864·789
3	1909·859
4	1432·394
6	954·930
10	572·957

Notes 1. Curves are described by the angle at the centre subtending an arc of 100 units
2. The deflection angle from the tangent for a chord of 100 units is half the degree of curvature
3. A 5° curve has a radius of 1146(1145·916) units
4. In the United Kingdom tables of preferred radii in degree terms are issued for road design

Fig. 6.3 Curve ranging − degree curves.

degree of curvature for the main curve. In towns they may be entirely transitional, i.e. from radius infinity on the straight to maximum curvature, then decreasing again to straight.

The curves used are the spiral and the parabola; though they can be set out from tables without knowledge of their geometry, an understanding of their properties is of assistance to the setting out engineer.

The Deflection Method

In this, the most common method, the points on the curve are ranged by setting up the theodolite in the line of the tangent at one of the tangent points, and turning off successive angles to points on the curve separated by the sub-chord distance in use. It is customary for these to be at 10 m intervals. This is not to say that they may not be at shorter distances; with sharp curves this may well be the case.

It is the convention that the chainage points at every sub-chain from the origin of the road construction contract are carried on round the curve so that the points to be marked will be at exact chainages, for example 3 + 10, 3 + 20, etc. Since curve lengths are seldom exact multiples of 100 m,

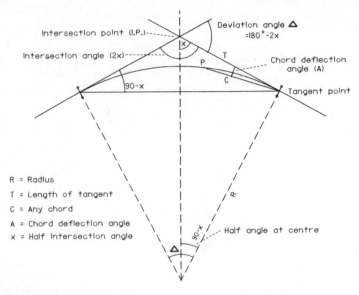

R = Radius
T = Length of tangent
C = Any chord
A = Chord deflection angle
x = Half intersection angle

Relationships

$$T = R \cot x = R \tan (90-x)$$

$$\sin A = \frac{C}{2R}, \quad R = \frac{C}{2 \sin A}$$

Length of curve

$$\text{Length} = \frac{\text{Angle at centre}}{360} \times 2\pi R = \frac{\Delta}{360} \times 2\pi R$$

Notes

1. Deviation angle Δ is twice the full chord deflection between tangent points $(90° - x)$
2. The angle at any point (P) on the curve between the two tangent points $= 90° + x$
3. Angle at centre = Deviation angle Δ

Fig. 6.4 Curve data.

and the chainage of the tangent points depends on their distance from the IP, it follows that the first and last sub-chords will be odd lengths. The initial and final deflections are, however, calculated by simple proportion:

$$\text{Sub-chord deflection} = \text{Chord deflection} \times \frac{\text{Sub-chord}}{\text{Chord}}$$

The setting out data may have been precalculated, but let us assume that they have not, and go through the complete process. The data required are as follows (see Fig. 6.5):

• The intersection angle of the tangents
• The radius of curve.

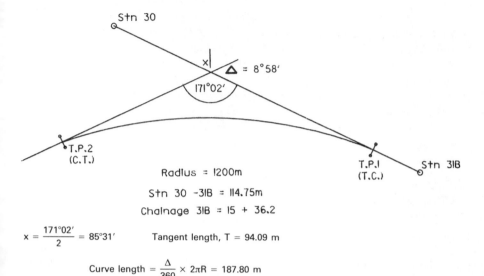

Stn 30

$\triangle$ = 8°58'

171°02'

T.P.2
(C.T.)

T.P.I
(T.C.)

Stn 31B

Radius = 1200m

Stn 30 -31B = 114.75m

Chainage 31B = 15 + 36.2

$x = \dfrac{171°02'}{2} = 85°31'$ Tangent length, T = 94.09 m

Curve length $= \dfrac{\Delta}{360} \times 2\pi R = 187.80$ m

Deflection angles (10 m chord)

$\sin A = \dfrac{C}{2R} = \dfrac{10}{2400} = 0.0041667$ radians, A = 14'19"

Initial sub-chord
Distance 31B to TP1 = 114.75 − T = 20.66
   Chainage of TP1 = 15 + 56.86
Nearest whole chord on curve is 15 + 60
   First sub-chord = 15 + 60 − 15 + 56.86 = 3.14

Sub-chord deflection $= \dfrac{3.14}{10} \times 14'19'' = 04'30''$

Final sub-chord
Chainage TP1 = 15 + 56.86
Curve length = 1 + 87.8
Chainage TP2 = 17 + 44.66

Final sub-chord = 4.66

Final sub-chord deflection $= \dfrac{4.66}{10} \times 14'19'' = 06'40''$

Fig. 6.5   Circular curve calculations.

From these two pieces of information from the plans can be calculated:

• The tangent length
• The length of the curve.

By setting back the calculated tangent length T from the intersection point (IP), the position of the first tangent point can be fixed. This is done by setting up a theodolite over the last chainage peg in the straight and taping back from the IP the distance T. The IP mark is then laid on and the TP peg set in exactly in the line of the tangent at its correct distance from the IP. Measure from the last chainage peg in the straight to derive the chainage of the tangent point.

The chainage of the further tangent point can now be found by adding the calculated curve length to the first TP chainage.

The final TP is set out in the same way as the first. Having now obtained the chainages of the TPs, the lengths of the initial and final sub-chords can be found, e.g.

Chainage TP 1	15 + 56.86	
Chainage next whole sub-chord	15 + 60	
Difference	3.14	= initial sub-chord

The next step is to prepare a table of deflections for each sub-chord peg on the curve. This is done by calculating the whole sub-chord deflection from the formula shown on Fig. 6.4, i.e.

$$\text{Sin } A = \frac{C}{2R}$$

Calculate the initial and final sub-chord deflections as a proportion of this figure.

As a check on the correctness of this work add up all the whole sub-chord deflections and the initial and final deflections. The result should be equal (+ or − a few seconds) to the total long chord deflection from one TP to the other (see worked example, Figs. 6.5 and 6.6).

If the TPs are intervisible a further check on the correctness of the work can be made, as follows:

(1) Set up the theodolite over TP 1 and lay at 0−180° in the line of the tangent. Either do this by setting 00°00′00″ on the plate and laying on the IP or set 180°00′00″ and lay back on some distant mark on the straight.
(2) Unclamp the top plate and lay on TP 2 and read the angle, which should agree within very close limits with the full deflection angle (90° − x). If the curve is left-handed, the reading will be an anti-clockwise one and must be subtracted from 360° to compare it.

With the table of chainages and deflections prepared, it is now possible to set out all the chainage pegs in the curve as follows:

(1) With the instrument set up correctly at TP 1 (0−180° in the tangent), unclamp the top plate and set on the initial sub-chord deflection angle. Direct the assistant with a peg into this line.

CHAINAGE	DEFLECTION			NEAREST 20"			L.H. DEFLECTION			CHORD
	°	'	"	°	'	"	°	'	"	
15 + 56.86	00	00	00	00	00	00	00	00	00	—
15 + 60	00	04	30	00	04	20	359	55	40	3.14
15 + 70	00	18	49	00	18	40	359	41	20	10
15 + 80	00	33	08	00	30	00	359	27	00	10
15 + 90	00	47	27	00	47	20	359	12	40	10
16 + 00	01	01	46	01	01	40	358	58	20	10
16 + 10	01	16	05	01	16	00	358	44	00	10
16 + 20	01	30	24	01	30	20	358	29	40	10
16 + 30	01	44	43	01	44	40	358	15	20	10
16 + 40	01	59	02	01	59	00	358	01	00	10
16 + 50	02	13	21	02	13	20	357	46	40	10
16 + 60	02	27	40	02	27	40	357	32	20	10
16 + 70	02	41	59	02	42	00	357	18	00	10
16 + 80	02	56	18	02	56	20	357	03	40	10
16 + 90	03	10	37	03	10	40	356	49	20	10
17 + 00	03	24	56	03	25	00	356	35	00	10
17 + 10	03	39	15	03	39	20	356	20	40	10
17 + 20	03	53	34	03	53	40	356	06	20	10
17 + 30	04	07	53	04	08	00	355	52	00	10
17 + 40	04	22	12	04	22	20	355	37	40	10
17 + 44.66	04	28	52	04	29	00	355	31	00	4.66

Closing error check 90° − x = 90° − 85° 31′
= 04° 29′
Discrepancy = − 8″

Fig. 6.6 Setting out table (20″ theodolite).

(2) The assistant positions a peg at the correct sub-chord distance from the TP and drives it in, watched by the engineer through the instrument to ensure that it stays in line and upright. When the peg is firm the assistant, using the tape again, scribes a line across the top of it,

and then places a mark on this line for the engineer to sight on. (A yellow pencil makes a good mark.) The assistant moves the mark as directed until it is exactly in line. A nail is then driven in, leaving about 4 mm protruding, and finally checked both for line and distance and marked with the correct chainage (Fig. 5.1).

(3) The next deflection angle is now set on the plate of the instrument and the assistant at full sub-chord distance (10 m) from the first peg is lined in as before. It is common practice when doing this kind of work to loop the ring of the tape over the nail in the peg and use only one assistant. If this is done, great care needs to be taken not to pull pegs and allowance must be made for the ring and the nail thickness if poor measurement is not to result. It is generally uneconomic to use only one person owing to the resulting slowness of the whole procedure.

(4) The process is repeated until the final sub-chord is reached. If the work has been done well the mis-close will be small (about 10 mm in 150 m). If larger than this a new peg should be set in and the mis-close measured so that the engineer can inspect it and be in a position to assess which kind of correction to make, linear or angular. The error should be eased out over a number of sub-chords depending on its size.

(5) Check back on reference object, i.e. IP or point on the straight, to see that the instrument has not moved off its original setting. If it has, the work may have to be repeated unless quick deflection checks show that the error occurred at a late stage and can be corrected easily. Great care in instrument handling is needed when constantly setting new angles and re-sighting, to ensure that the original setting is not disturbed or the wrong tangent screw moved. It is good practice to check back on the RO (reference object) from time to time.

In all these operations it is far better to be methodical and deliberate and to get the right answer the first time. Haste invariably wastes more time than it saves in detailed work.

When checking possible mis-closes it may be obvious that the error is either angular or linear and this should be taken into account when correcting. It is possible that errors may have occurred in setting out the TPs or in the calculation of the curve length. Careful examination of all data should be done in the case of a large mis-close before the curve setting is corrected. However, if the work has been done well and the checks described done at the right state, large errors are unlikely to occur at the end.

The error caused by measuring chords of 10 m rather than arcs can be ignored for all practical purposes, even on quite short radius curves.

Station Change

In the foregoing procedure it was assumed that the whole curve was set out from the tangent point. In practice this seldom occurs, or is possible, for visibility reasons alone. What to do when changing the position of the theodolite is set out in the following paragraphs, and includes situations in which the intersection point itself may be inaccessible.

Single Change of Theodolite Station

When it is no longer possible to set out further sub-chords from the TP, proceed as follows:

(1) Re-set the instrument over the last chainage peg fixed.
(2) Lay back on the TP with 180° reading on the plate.
(3) Unclamp the top plate and swing on to TP 2 as a check; the reading should be equal to the full deflection angle $(90° - x)$.
(4) Using the table of deflections previously calculated, set on the deflection for the next sub-chord exactly as though at TP 1 (Fig. 6.7).

Further Change of Theodolite Station

When making a further instrument move, proceed as follows:

(1) Lay back on the furthest visible station with 180° plus the original deflection to that station on the plate (Fig. 6.7).
(2) Unclamp the top plate; the reading to TP 2 will be the same as before, the full deflection angle $(90° - x)$.
(3) Set on subsequent deflections exactly as before, as though at TP 1.

Inaccessible Intersection Points

It has been assumed that the intersection point is accessible, but this may not always be the case; for example, a curve rounding a headland may have the intersection of the tangents in the sea. However, if the angular relationship of the two straights can be measured then the intersection angle can be deduced. To do this, proceed as follows:

(1) Find two points, A and B, on the tangents which are intervisible (Fig. 6.7).
(2) Carefully establish a peg at each place which is correctly in the line of the tangent; for effective results it should be lined in with an instrument.

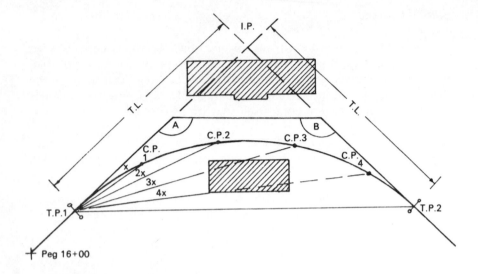

1. Establish points A and B on respective tangents
2. Measure angles A and B and distance AB
3. Intersection angle = $\hat{A} + \hat{B} - 180°$
4. Calculate tangent length
5. Solve triangle A.IP.B
6. Set in T.P.1 @T.L. — Distance A-I.P. from A
7. Set in T.P.2 @T.L. — Distance I.P.-B from B
8. Measure distance T.P.1 to peg 16+00 to give chainage of T.P.1
9. Calculate chord and subchord deflections for setting out
10. Set out all visible chord points. In this example C.P.1 and C.P.2 (C.P.3 is not visible)
11. Change station with theodolite to C.P.2

Procedure at intermediate station C.P.2
1. Lay back on T.P.1 with 180° on the plate, unclamp top plate. Turn to deflection for C.P.3 as if at T.P.1 (3x in this example)
2. Check total deflection to T.P.2 if visible

Procedure at further change station (C.P.3)
1. Lay back on, say, C.P.1 with 180°+reading to C.P.1 (180+x in this example). Unclamp top plate, turn to next deflection (C.P.4) as if at T.P.1 (4x in this example)
2. Check on T.P.2 as above.

Fig. 6.7 Setting out a curve where the intersection point is inaccessible and station changes are required.

(3) At both A and B, measure the external angles of the triangle A.IP.B.

    Angle A = TP 1.A.B.

    Angle B = TP 2.B.A.

The sum of these two angles minus 180° will equal the intersection angle.

(4) Measure the distance AB.

(5) From this information it is now possible to calculate the sides of the triangle A.IP.B. Use the sin rule:

$$\frac{a}{\sin A} = \frac{b}{\sin B} = \frac{c}{\sin C}$$

$$\frac{AB}{\sin \text{intersection } \angle} = \frac{IP - A}{\sin 180 - B} = \frac{IP - B}{\sin 180 - A}$$

$$IP - A = \frac{AB \sin B^*}{\sin \text{intersection } \angle}$$

(* sin B = sin 180 − B. This avoids a subtraction operation and saves time)

The tangent length T is known from the original data ($T = R \cotan \frac{1}{2}$ intersection angle) so that the relationship of the TPs with stations A and B is now fixed; e.g. suppose the tangent length to be 314.76 and the side IP − A to be 232.45, then obviously the difference 82.31 is the distance between Station A and TP 1, which can now be set out by direct measure from A.

It is, of course, possible that two points A and B which are intervisible could not be found, but the method is not affected. It will be necessary to traverse between them and establish their linear distance and the angular relationship of the tangents by computing the coordinates of one in terms of the other.

In certain cases it may happen that the tangents of a road centre line have been fixed in relation to original survey marks and coordinates for points on them are available. Alternatively certain key points may be fixed in terms of National Grid coordinates. In either case, it is quite a simple task to compute intersection angles to close limits so that the curve calculated from such data will close no less accurately than one where the angle is observed. The position of the tangent points can be similarly fixed by calculating their position in coordinate terms.

When establishing the exact position of an IP if the straights have been fixed at other points, great care should be taken, and it is necessary to line in the position of intersection of tangents using a theodolite if reasonable results are to be obtained (Fig. 6.8).

Curve Ranging by Linear Measure

There are two common methods of setting out curves by using linear measurement:

(1) The 'offset from the chord produced' method
(2) The mid-ordinate method

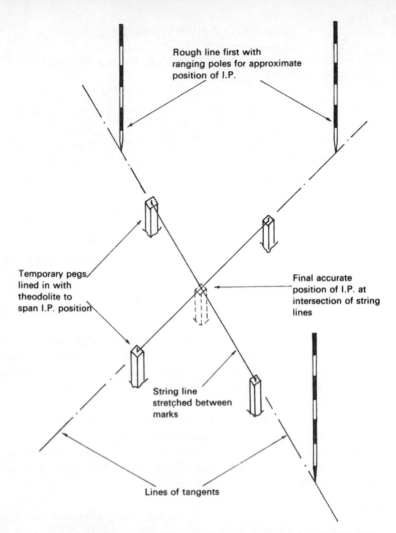

Rough line first with
ranging poles for approximate
position of I.P.

Temporary pegs
lined in with
theodolite to
span I.P. position

Final accurate
position of I.P. at
intersection of string
lines

String line
stretched between
marks

Lines of tangents

Notes 1. Temporary pegs should be at the same height to ensure physical intersection
          of string lines
       2. Accuracy of operation will depend on requirement, but a good curve
          mis-close requires an accurate position and intersection angle

Fig. 6.8 Curve ranging – setting IP pegs from tangents.

In the 'chord produced' method the offset to the first point of the curve is made from the tangent. Subsequent offsets are made from the line of each successive chord produced an equal distance forward. It is convenient to choose a chord length which is half a standard tape length, e.g. 15 m for a 30 m tape. There is nothing to prevent shorter chords being used, however.

The first offset is calculated from the formula $\dfrac{C^2}{2R}$

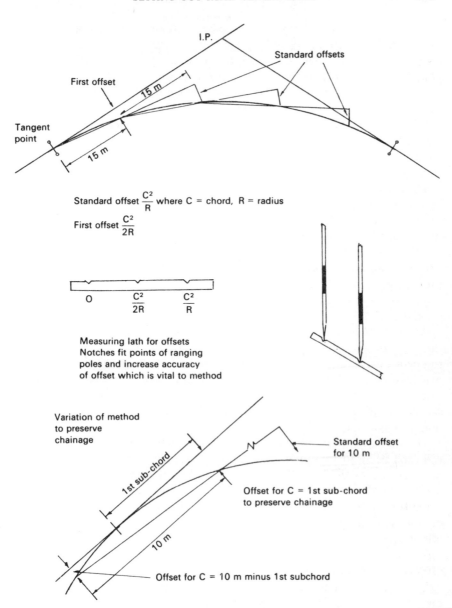

Standard offset $\dfrac{C^2}{R}$ where C = chord, R = radius

First offset $\dfrac{C^2}{2R}$

Fig. 6.9 Curve ranging by linear measurement.

where C is the chord length chosen and R the curve radius. The subsequent offset from the chord produced is twice this figure.

For most radii the two offsets can be marked accurately on a light piece of wood, and a notch made opposite each length and the zero to assist in maintaining accuracy of measure with taping arrows or the points of ranging poles.

Procedure is as follows: with the assistant holding the zero end of the tape at the TP, a ranging pole is lined in on the tangent distant to the chord length chosen. Using the lath to measure the offset, another pole is planted on the curve at the same chord distance from TP.

The engineer now walks forward reeling out the tape for a further chord length, and lines in on the line first chord point − TP, planting the ranging pole at the chord produced distance. The assistant now walks forward to the first chord point and holds the tape at the point while the engineer offsets a further ranging pole on the curve, this time using the standard offset distance.

The engineer now walks forward a further chord length, dragging the tape until the zero end is at the assistant's position. The assistant holds the zero end of the tape and the engineer lines in on the new chord and offsets as before. This procedure is repeated chord length by chord length until the curve is complete.

Care in lining in and measuring the offset is required, as any error in this respect is doubled for each chord. It is better to use ranging poles originally so that lining in is quick and the whole curve can be visually checked.

At the final mark the chord length will either fall short of or overshoot the TP (unless an exact fraction of the curve length has been chosen) and this will enable any misclose to be judged. Any error should be eased over a number of chord lengths, halving the correction each time. For example, if the last pole had to be moved in 100 mm the next to last should be moved in 50 m and the next 25 mm and so on. The error in length is not likely to be great.

The method is suitable for quick ranging of lines for site clearance, but can also be used for quite short radius curves where centre points are inaccessible. With short chord lengths and stringlines over iron pins or arrows for the lining in, very reasonable results can be obtained, with care. It is difficult to put in pegs at exact chainages by this method for obvious reasons. It can be done, however, by calculating the offset for the initial sub-chord and setting out this point, then calculating a similar offset for a length which is the difference between the standard chord (10 m) and the initial sub-chord and setting out this point back from the TP, not forward. This produces a part of the curve which is touching the tangent at the TP. Using these two points, which are a whole chord length apart, the method can be used as before; each successive point in this case will now be at its correct chainage (Fig. 6.9).

The offset method (sometimes called the kerblayers' offset) is useful for short radius curves where the long chord between tangent points is of reasonable taping distance, one or two tape lengths.

The offset, M, is calculated and this distance offset from the mid point

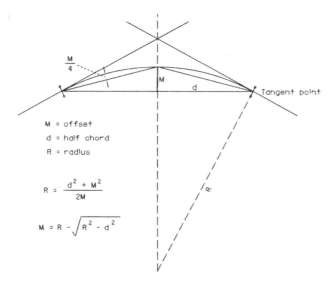

M = offset
d = half chord
R = radius

$$R = \frac{d^2 + M^2}{2M}$$

$$M = R - \sqrt{R^2 - d^2}$$

Subsequent offsets for each successive halving of the chord length are each one quarter of the previous one.

The method is much used by kerb layers for setting out kerb pins between engineer's pegs.

Fig. 6.10 Curve ranging by linear measurement — the offset method.

of the long chord, thus marking the centre of the curved arc. The offsets for successive half chords, quarter chords, etc., for all practical purposes are equal to $\frac{M}{4}$, $\frac{M}{16}$, $\frac{M}{64}$, etc.

A kerblayer given the TPs and the offset will lay a satisfactory curve by this method, halving successive chords and quartering offsets (Fig. 6.10).

Transition Curves

The curves most commonly used to provide transition from the straight to a planned circular curve on roads are the clothoid (sometimes called Euler's spiral) and the parabola. They both start from a radius of infinity and gradually sharpen (decrease in radius). In their first portions the curves are very similar in shape. The spiral, however, continues to decrease in radius while the parabola does not. For motorway work the spiral is generally used, but sometimes on low speed urban roads and junctions the parabola is more useful.

Both curves can be set out by deflection angles in the same way as circular curves, but since the curve is sharpening there is no simple sub-chord deflection to use for each successive point on the curve.

This can be overcome in various ways, of which the two most simple are the use of convenient formulae and recourse to tables of prepared data. Before metrication, the most widely used tables were *Highway Spirals and Superelevation Tables*, H. Criswell, published by The Carriers Publishing Co. Ltd.

These tables were calculated for certain designed speeds of traffic and provided all the necessary data for setting out in imperial units. They were accepted as standard by the then Ministry of Transport. Since metrication, the County Surveyors' Society realised the need for similar tables and compiled them using metric units in very much the same form as Criswell's *Tables*. These have also received official recognition.

There are certain terms and relationships which must be understood to use the tables and to set in the necessary centre line pegs (Fig. 6.11). The start peg, for example, when coming off the straight, is no longer marked TC as with a circular curve, but TS (tangent to spiral). The point of change from straight to curve is also moved back along the tangent by the insertion of the spiral lead-in to the main curve. Consequently the length *T* (tangent length in circular curves) will be increased by an amount denoted as *C*.

This also has the effect of moving the circular part of the curve towards the centre by an amount known as *S* (shift). The radius of the new circular curve is derived from the application of the shift to the original circular curve radius; this new radius is denoted by *R*. Both figures are tabulated for all lengths of transition spiral.

Tangent length *T* is originally calculated by using *R* + *S* values and the total apex distance from IP to TS is obtained by adding the value for *C* from the tables. This value is looked up against the length of spiral in use in the column headed 'Spiral Length, *L*'.

The tables show deflection angles from the origin (TS) for every 1 m, 2 m or 5 m of transition spiral depending on the degree of curvature.

Also tabulated is a figure called the 'back angle' for every chord point. Using this value on the plate of the theodolite with the instrument at the chord point, and laid on the origin (TS), the 0−180° line will be the tangent to the spiral at that point. This angle is usually twice the value of the deflection angle, but there are small corrections to be made in long transitions; these are tabulated separately.

Other items tabulated are 'angle consumed by the spiral', which means the angle at the centre subtended by the spiral at that point, and the 'degree of curve' at any point which is the radius of the spiral at any point in degree terms. When changing from spiral to circular curve it is useful, in providing deflection values for setting out further chords on the circular part, to know the coordinates *X* and *Y* for each point, the *X* value being

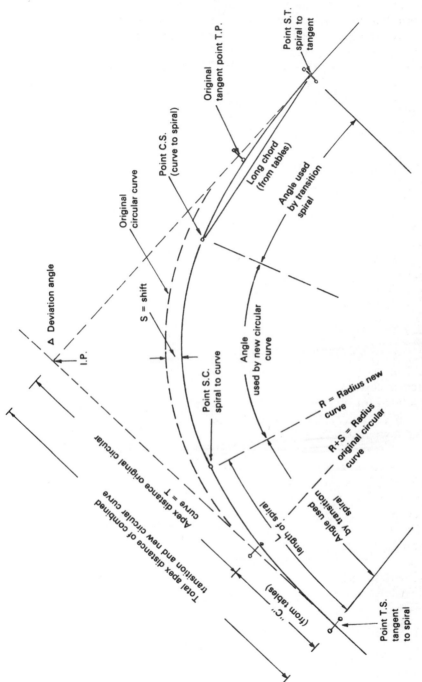

Fig. 6.11 Transition curves – terms and relationships.

Table 6.1 *Highway Transition Curve Tables (Metric)* – Table 9, reproduced by courtesy of the Country Surveyors' Society.

RADIUS R (METRES)	DEGREE OF CURVE D (° ' ")	SPIRAL LENGTH L (METRES)	ANGLE CONSUMED (° ' ")	SHIFT S (METRES)	R + S / R (METRES)	C (METRES)	LONG CHORD (METRES)	CO-ORDINATES X (METRES)	CO-ORDINATES Y (METRES)	DEFLECTION ANGLE FROM ORIGIN (° ' ")	BACK ANGLE TO ORIGIN (° ' ")
2864.7890	2 0 0.0	5.00	0 3 0	0.0004	2864.7893	2.5000	5.0000	5.0000	0.0015	0 1 0.0	0 2 0.0
1432.3945	4 0 0.0	10.00	0 12 0	0.0029	1432.3974	5.0000	10.0000	10.0000	0.0116	0 4 0.0	0 8 0.0
954.9297	6 0 0.0	15.00	0 27 0	0.0098	954.9395	7.5000	15.0000	14.9999	0.0393	0 9 0.0	0 18 0.0
716.1972	8 0 0.0	20.00	0 48 0	0.0233	716.2205	9.9999	19.9998	19.9996	0.0931	0 16 0.0	0 32 0.1
572.9578	10 0 0.0	25.00	1 15 0	0.0455	573.0032	12.4998	24.9995	24.9988	0.1818	0 25 0.0	0 50 0.2
477.4648	12 0 0.0	30.00	1 48 0	0.0785	477.5434	14.9996	29.9887	29.9870	0.3141	0 36 0.0	1 12 0.7
409.2555	14 0 0.0	35.00	2 27 0	0.1247	409.3803	17.4988	34.9960	34.9936	0.4988	0 49 0.0	1 38 1.4
358.0986	16 0 0.0	40.00	3 12 0	0.1861	358.2248	19.9999	39.9925	39.9875	0.7445	1 3 59.9	2 8 1.2
318.3099	18 0 0.0	45.00	4 3 0	0.2650	318.5749	22.4970	44.9900	44.9775	1.0599	1 20 0.2	2 42 0.2
286.4789	20 0 0.0	50.00	5 0 0	0.3635	286.8424	24.9949	49.9610	49.9610	1.4537	1 39 59.6	3 20 0.7
260.4354	22 0 0.0	55.00	6 3 0	0.4838	260.9191	27.4913	54.9387	54.9387	1.9343	2 0 59.8	4 2 1.3
238.7324	24 0 0.0	60.00	7 12 0	0.6280	239.3604	29.9860	59.9579	59.9053	2.5104	2 23 58.8	4 48 2.5
220.3684	26 0 0.0	65.00	8 27 0	0.7982	221.1666	32.4765	64.9372	64.8588	3.1904	2 48 58.1	5 38 1.9
204.6278	28 0 0.0	70.00	9 48 0	0.9987	205.0245	34.9659	69.9090	69.7955	3.9887	3 15 57.1	6 32 13.2
190.9859	30 0 0.0	75.00	11 15 0	1.2255	191.2114	37.4519	74.8716	74.7114	4.8952	3 44 55.6	7 31 24.8
179.0493	32 0 0.0	80.00	12 48 0	1.4867	180.5360	39.9335	79.8227	79.6017	5.9362	4 15 53.5	8 32 35.2
168.5170	34 0 0.0	85.00	14 27 0	1.7824	170.2994	42.4101	84.7600	84.4609	7.1133	4 48 50.6	9 38 9.4
159.1549	36 0 0.0	90.00	16 12 0	2.1452	161.2695	44.8804	89.6806	89.2832	8.4340	5 23 46.8	10 48 44.0
150.7784	38 0 0.0	95.00	18 3 0	2.4944	153.2935	47.3416	94.5816	94.0816	9.9005	6 0 41.8	12 2 57.4
143.2394	40 0 0.0	100.00	20 0 0	2.8903	146.1357	49.7976	99.4595	98.7884	11.5337	6 39 35.2	13 20 14.4
136.4185	42 0 0.0	105.00	22 3 0	3.3496	139.7682	52.2419	104.1555	103.4555	13.3328	7 20 26.7	14 42 35.1
130.2177	44 0 0.0	110.00	24 12 0	3.8471	134.0648	54.6746	109.1303	108.0538	15.2007	8 3 16.0	16 8 44.6
124.5560	46 0 0.0	115.00	26 27 0	4.3905	129.3966	57.0939	113.9144	112.5733	17.2573	8 48 4.6	17 38 31.3
119.3662	48 0 0.0	120.00	28 48 0	4.9814	124.3476	59.4982	118.6579	117.0033	19.7662	9 34 45.6	19 12 8.5
114.5916	50 0 0.0	125.00	31 15 0	5.6214	120.2130	61.8854	123.1551	121.3324	22.2473	10 23 24.9	20 51 35.1
110.1842	52 0 0.0	130.00	33 48 0	6.3110	115.4962	64.2533	128.0004	125.5482	24.9348	11 13 59.5	22 34 46.3
106.1033	54 0 0.0	135.00	36 27 0	7.0544	113.1577	66.5996	132.5873	129.6378	27.8108	12 6 28.7	24 20 31.3
102.3139	56 0 0.0	140.00	39 12 0	7.8499	110.1637	68.9219	137.1091	133.5873	30.8762	13 0 51.5	26 11 8.5
98.7858	58 0 0.0	145.00	42 3 0	8.6994	107.4853	71.2176	141.1585	137.1823	34.1308	13 57 6.0	28 5 53.1
95.4930	60 0 0.0	150.00	45 0 0	9.6640	105.1570	73.4840	145.5278	141.0078	37.5732	14 55 13.7	30 4 46.3
92.4125	62 0 0.0	155.00	48 3 0	10.5642	102.9768	75.7182	150.2090	144.6481	41.2007	15 55 10.8	32 7 49.2
89.5247	64 0 0.0	160.00	51 12 0	11.5807	101.1053	77.9172	154.3933	147.6872	45.0088	16 56 56.5	34 15 3.5
86.8118	66 0 0.0	165.00	54 27 0	12.6536	99.4654	80.0778	158.4717	150.7086	48.9918	18 0 29.3	36 26 30.7
83.2585	68 0 0.0	170.00	57 48 0	13.7850	97.0435	82.1967	162.4347	153.6937	53.1421	19 5 47.4	38 42 12.6
81.8511	70 0 0.0	175.00	61 0 0	14.9607	96.8118	84.2707	166.2722	156.0318	57.4504	20 12 48.7	41 2 11.3

RADIUS R (METRES)	DEGREE OF CURVE D (° ' ")	SPIRAL LENGTH L (METRES)	ANGLE CONSUMED (° ' ")	SHIFT S (METRES)	R + S / R (METRES)	C (METRES)	LONG CHORD (METRES)	CO-ORDINATES X (METRES)	CO-ORDINATES Y (METRES)	DEFLECTION ANGLE FROM ORIGIN (° ' ")	BACK ANGLE TO ORIGIN (° ' ")
229.1831	25 0 0.0	62.50	7 48 0	0.7097	229.8928	31.2306	62.4484	62.3859	2.8309	2 36 13.5	5 12 31.5
190.9859	30 0 0.0	75.00	11 15 0	1.2255	192.2114	37.4519	74.3716	74.7114	4.8952	3 46 5.6	7 30 41.4
163.7022	35 0 0.0	87.50	15 19 0	1.9438	165.6460	43.6400	86.7226	86.8771	7.7552	5 6 53.2	10 13 41.1
143.2394	40 0 0.0	100.00	20 0 0	2.8963	146.1357	49.7976	98.4595	98.7884	11.5337	6 39 35.2	13 8 24.8
127.3240	45 0 0.0	112.50	25 18 45.0	4.1150	131.4370	55.8860	111.5271	110.3260	16.3375	8 25 24.6	16 53 20.4

GAIN OF ACCN. m/s³	0.30	0.45	0.60
SPEED VALUE km/h	58.5	67.0	73.7

INCREASE IN DEGREE OF CURVE PER METRE = D/L = 0° 24' 0.0"

RL CONSTANT = 14323.945

DEGREE OF CURVATURE BASED ON 100 m. STANDARD ARC

the distance along the tangent, and the Y the offset. They can be used for setting in points by linear measure in certain circumstances.

The formulae for both the transition spiral and the cubic parabola when calculating ordinates are identical when small angles of deflection only are considered, e.g.

$$\text{(the offset) } Y = \frac{X^3}{6RL}$$

where R and L are radius and length of spiral respectively. In terms of ordinates the formula for the cubic parabola is

$$Y = aX^3$$

where a is a constant for the chosen curve.

The spiral formula is $L = a\sqrt{\phi}$ where a is a constant and ϕ is the angle (in circular measure, radians) which the tangent at any point makes with the original tangent. The constant a in use is $\sqrt{2RL}$ where R and L are radius and length; thus RL is constant, and this figure will be found in the tables and can be used to find either quantity when the other is known.

Use of the tables and practice in calculating the data for transition curves will promote an understanding of these figures for anyone engaged in setting out, whatever his or her mathematical training.

Though in many cases where transition curves are to be used the data for setting out may have been calculated at the design stage, the setting out engineer will gain better understanding of his job and be able to cope with discrepancies as they occur in practice, if calculations are done. Often it is necessary to measure or re-measure intersection angles on the ground, and reconciling the design measurements with the ground may make some re-calculation necessary. Accordingly, the method of calculating setting out data for a simple transition curve is set out and a worked example shown with Fig. 6.14.

The requirements are similar to those for a circular curve with the additions already described:

- The intersection angle
- The proposed radius for the curve
- The nominal speed for which the road is designed.

From this information can be calculated:

- The tangent length for the original curve
- The total length of the new curve and the spiral addition.

Proceed as follows:

(1) Measure the intersection angle (unless already known).
(2) Calculate T (tangent length) for original curve radius. Use $R + S$ value for radius.

(3) Look up appropriate table for design speed (say 67 km/h).

(4) Against the nearest value to the original curve radius in $R + S$ column, take out C for addition to T, and note spiral length and angle consumed.

(5) Add C to T for total apex distance IP to TS.

(6) Deduce (or measure) chainage of TS.

(7) Subtract twice the angle consumed from the deviation angle (delta), thus giving the remaining angle at the centre consumed by the circular arc.

(8) Add length of spiral to chainage of TS, giving chainage of point SC (spiral to curve).

(9) Calculate length of circular curve, using R.

(10) Add this length to chainage of SC, thus giving chainage of point CS (curve to spiral).

(11) Add length of spiral to chainage of point CS, thus obtaining chainage of point ST (spiral to tangent).

(12) Prepare tables of deflection angles against chainage for all points on curve including initial and final sub-chords for both spiral and circular curve. (See worked example and Figs. 6.12 and 6.13.)

The setting out procedure is similar to that for circular curves and is made difficult in the same way by the chainage convention. While it is possible to extract the deflection angles direct from the table, they must be adjusted to suit the differing lengths of the spiral from the origin, caused by the odd TS chainage and the initial and final sub-chords. Both sets of tables show methods of calculating deflection for odd chord lengths and have tables of corrections for approximations made. However, up to deflections of about 7° the maximum correction is not larger than 30″ of arc, so that unless the transition is very long, or the radius very sharp, the methods shown in the figures will generally suffice without too much detailed correction. A balance between theoretical correctness and practical site dimensions must be struck.

Until a high degree of skill with a theodolite and linear measurement is attained, it is useful to divide the setting out into parts. First, points TS and ST are set out in the tangents; then, by using the full spiral deflection and the long chord distance from the tables, points SC and CS are set in.

This will have the effect of revealing any initial errors, and will confine any subsequent ones on the intermediate chord points to small values. A check can be made by setting up the instrument at SC, laying back on TS with 180° minus the back angle on the plate, thus setting 0–180° tangent to both spiral and circular curve at this point, then checking angle to point CS. This should equal the full deflection (90° minus $\frac{1}{2}$ intersection angle) for the circular curve.

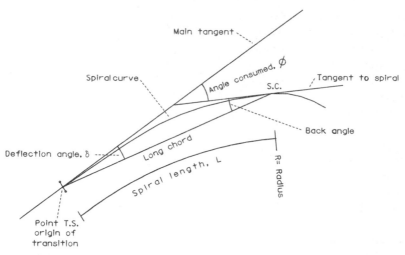

ϕ is the angle between the main tangent and the tangent to the spiral at any point.

Its value (in radians) $= \dfrac{L}{2R}$

where L = Long chord and R = Radius at any point.

δ is the deflection angle from origin TS and is equal to $\dfrac{\phi}{3}$ for small angles.

Obviously the back angle $= 2\phi$

Deflection angles to any point on the spiral can be calculated from the relationship $\left(\dfrac{l}{L}\right)^2 \times \dfrac{\phi}{3}$

where l = length to intermediate point.
Practical for angles up to 7°.

Examples

$L = 50$ m	Deflection	$= 01°40'00''$
67 km/h Table 6.1	Back angle	$= 03°20'00''$
	Sum $= \phi$	$= 05°00'00''$

Therefore deflection for first 10 m $= \left(\dfrac{10}{50}\right)^2 \times \dfrac{\phi}{3} = 00°04'00''$

Check deflection for 10 m from Table 6.1 $= 00°04'00''$

Alternatively $\phi = \dfrac{L}{2R} = \dfrac{10}{2 \times 1432.3945} = 0.00349066$ radians $= 00°12'00''$

$\delta = \dfrac{\phi}{3} = 00°04'00''$ as before

Fig. 6.12 Transition curves – spiral data.

The intersection angle is obtained by subtracting the angle consumed by the circular curve from 180° (Fig. 6.11). Small angular errors will occur but they should not be significant in terms of positional error.

In practice it is often not possible to set the entire transition from the

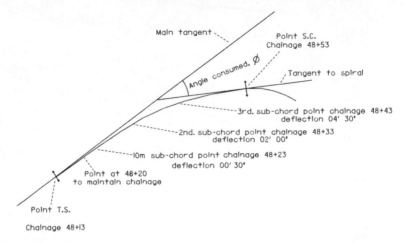

The convention that chainages are in multiples or sub-multiples of 100 m provokes certain difficulties for setting out. Curve tables are so arranged that deflections can be read straight from them for sub-chords of 1, 2 and 5 m without the need for calculations. This will, however, produce centre line pegs at odd chainages, e.g. 48 + 23. This problem can be overcome:

Using the relationships $\delta = \dfrac{\phi}{3}$ and $\left(\dfrac{l}{L}\right)^2 \times \dfrac{\phi}{3}$ = deflection for sub-chord length l, calculate deflections for odd chord lengths, e.g., in example 7, 17 and 27 m to maintain conventional chainages of 48 + 20, 48 + 30, etc.

Example: for 67 km/h transition 40 m long

L = 40 m   Deflection = 01°04′00″
                    Back angle = 02°08′00″

ϕ = 03°12′00″

$$\delta \text{ for 7 m} = \left(\frac{7}{40}\right)^2 \times \frac{03°12′00″}{3} = 01′58″$$

$$\delta \text{ for 17 m} = \left(\frac{17}{40}\right)^2 \times \frac{03°12′00″}{3} = 11′34″$$

Fig. 6.13 Transition curves − setting out.

origin TS and, as with circular curves, the theodolite must be moved round the curve. The methods used when setting out circular curves will not suffice in this case, but there are a number of ways of overcoming the problem (Figs. 6.14, 6.15 and 6.16).

Setting out by the use of coordinates for short transitions is not to be despised, using the long chord as check if the ground is suitable. Examination of the tables will show when the offsets are too large. It is, of course, usually necessary to line in points on the tangent at the X ordinate distances with an instrument, but this can be done with taping arrows for short distances. Two tapes are best used, 100 m for the X ordinates and long chord distances and 30 m for the offsets (Y ordinates).

As in all setting out work, it is vital to be methodical and neat in the preparation of data in the field book. No figure should be anonymous,

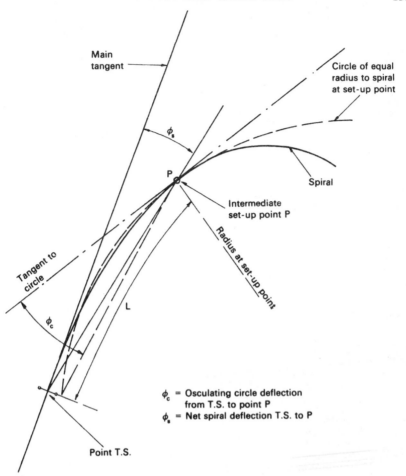

Fig. 6.14 Transition curves — setting out from intermediate point — osculating circle method.

and tabulation should be done wherever possible. Much time will inevitably be saved.

Calculations for Insertion of a Transition into an Existing curve

Figure 6.5 shows an example of a circular curve into which a 67 km/h transition is to be inserted.

Present curve data: Radius 1200 m. Intersection angle 171°02′. Chainage TC 15 + 56.86, CT 17 + 44.66.

(1) The sample table on p. 150 shows 1432.3945 as the nearest tabulated radius to the existing curve. Degree of curvature is 4°00′ and length of spiral (*L*) 10 m.

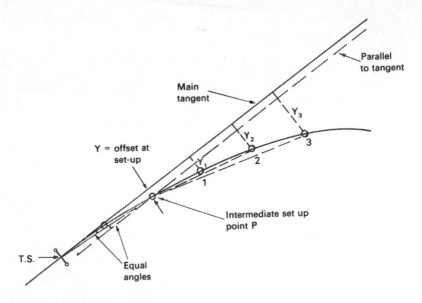

Notes   1. Lay 0−180° parallel to main tangent at set-up by laying back on T.S. (or last station) at 180°+deflection angle to set-up point (minus for L.H. curves)
        2. If last station sighted not T.S. use 180°+deflection to T.S.+deflection to sighted point
        3. Calculate further deflections by sin rule using Y_1, Y_2 and Y_3 ordinates less Y the ordinate at P as one side and chord lengths as the other side of the right angled triangles formed at points
        4. The deflection to point 3, for example, will have as its sin $\dfrac{Y_3-Y}{3C}$ where C = chord length
        5. Alternatively, when Y_1, Y_2 or Y_3 is below 30 metres set out by ordinates from the tangent

Fig. 6.15 Transition curves − setting out from intermediate point − another method.

(2) The correct length can be obtained as follows:

Divide the constant (RL) for the transition by the radius required. In this case:

$$\frac{14323.945}{1200} = 11.937$$

(3) The angle consumed at the centre for an L of 10 m is 00°12′00″.

There is an extra total length of 1.937 in the selected length of 11.937. This will increase the angle consumed by:

$$\frac{1.937}{1200} = 0.00161417 \text{ radians} = 05'33''$$

The original angle at the centre was 08°58′00″ (the deviation angle), so the new angle becomes

$$08°58′00″ - (24′00″ + 11′06″) = 08°22′54″$$

(4) The new length of circular curve for this angle is given by radius × angle (in radians).

$$1200 × 0.14628768 = 175.545$$

The effect of inserting the spiral 'lead in' is to increase the length of the curved portion of the alignment by one spiral length. The new chainage of the point of deviation from the straight (TS) will be at a different apex distance from the IP by the amount C adjusted for the spiral length chosen; in this case 5.969. Note that in this table at these radii, C is equal generally to half the spiral length.

(5) A new table of chainages should now be compiled so that the deflection angles for the new combined curve can be calculated:

Original chainage TC	15 + 56.86
Less C (from table)	0 + 5.969
New chainage, Point TS	15 + 50.891
Add L	0 + 11.937
New chainage, point SC	15 + 62.828
Add new curve length	1 + 75.545
New chainage, point CS	17 + 38.373
Add L	0 + 11.937
New chainage, point ST	17 + 50.31
Check. Deduct C	0 + 5.969
Compare old chainage CT	17 + 44.341
	17 + 44.66
Difference:	0 + 0.319

The rather large discrepancy is due to the fact that the selected radius of 1200 m was not very close to any of the tabulated radii. Therefore the calculations, particularly of the curve length, are rather poor. There are a number of possible solutions to this problem. If only tables are to be used then it is recommended that a radius value close to the tabulated values, or preferably a tabulated value itself, is used for the curve. However, if the already chosen value for the radius is to be used, then the discrepancy, if not too large, can be eased out in the setting out process. For a better solution the formulae listed at the beginning of the tables should be used instead of the tables themselves. With a small calculator this should not pose a problem.

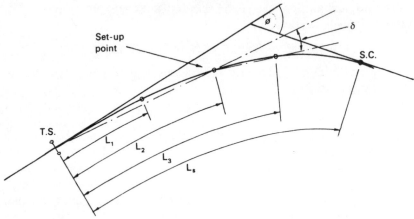

This method can be used at any point distant L_2 from origin with instrument laid on any previous point distant L_1 from origin (180° on plate)

Angle δ is the deflection from 0° to a point distant L_3 from origin

Calculate δ from the formula $\delta = \dfrac{(L_1+L_2+L_3)\,(L_3-L_1)}{L^2} \times \dfrac{\theta_s}{3}$

When the instrument is laid back on point T.S. as the back station L_1 and L_2 have the same value

Fig. 6.16 Transition curves – setting out from intermediate point – alternative method for angles up to about 7°.

(6) An immediate check on calculation can be done by comparison of the end results of a new deflection table with the angle consumed by the circular curve.

The calculation for the new full deflection is given by:

$$\frac{\text{Length of Curve}}{\text{Sub-chord}} \times \text{Sub-chord deflection}$$

$$= \frac{175.545}{10} \times 14'19'' \qquad\qquad \text{(Fig. 6.5)}$$

$$= 04°\ 11'19''$$

Compare this with half the angle at the centre:

$$08°22'54'' \div 2 = 04°11'27''$$

(7) The total deflection for the combined curve must now be calculated by arriving at the spiral deflections for $L = 11.937$ This is given by the expression from Fig. 6.12:

$$\left(\frac{\text{length from origin}}{\text{length of spiral (from tables)}}\right)^2 \times \frac{\theta_s}{2}$$

θ_s (theta spiral)

$\quad$ = Back angle + deflection

$$= \left(\frac{11 \cdot 937}{10}\right)^2 \times \frac{12}{3}(\text{min}) = 5.7 \text{ min} = 05'42''$$

Considering the curve from the point of view of setting out right-handed, the back angle at Point SC will be twice this figure and θ_s three times. The difference of inclination from the original of the new tangent at this point will therefore be $17'06''$.

The original full deflection for the circular curve was:

$$4°28'52''$$
$$\text{Deduct } \theta_s \quad 0°17'06''$$
$$\overline{4°11'46''}$$

The reasons for the rather large discrepancy are the same as those give in (5).

It now merely remains to compile the deflection table correctly to ensure that the new transition curve will close. The process is rendered laborious by the necessity of preserving the chainage points round the curve.

It is usually not advisable to make a complete table to include both transition and circular curves, but rather to set out the transitions separately to points SC and CS. This set can be done either by offset from the tangent or by long chord deflection and distance. Establishing these points gives closure checks for both curves for the sub-chord deflections.

This simple exercise will give a good understanding of the practical steps required when pre-calculated data must be amended on site.

Calculation of Instrument Readings from an Intermediate Point
(method using the osculating circle as a circular tangent – Fig. 6.14)

(1) Look up degree of curvature of spiral at set up point, P.
(2) Calculate the deflection from TS for P on a circle of this radius (ϕ_c) using the formula for circular curves

$$\sin \phi_c = \frac{\text{chord}}{2R} = \frac{L}{2R} \text{ or } \phi_c = \frac{D}{2} = (\text{for 100 chord})$$

(3) Subtract the net spiral deflection (ϕ_s) to P from the figure obtained.
(4) Set the result ($\phi_c - \phi_s$) + 180° (180° – ($\phi_c - \phi_s$) for left-hand curves) on the instrument and lay back on TS.

(5) Release top plate and set further deflections for succeeding points, as follows: to the circle deflection for the next sub-chord add the *first* spiral deflection from the table.
(6) For the next sub-chord set twice the circle deflection plus the *second* spiral deflection from the table and so on.

Calculations for a 67 km/h curve ($L = 100$ m) are as follows.
Point P is at 50 m from origin (TS).
Curve radius at this point $286.4789 = 20°00'$ in degree terms.

$$\therefore \text{ deflection for 100 m chord} = \frac{D}{2} = 10°00'00''$$

$$(\phi_c) \text{ deflection for 50 m} = \frac{50}{100} \times \frac{D}{2} = 5° \; 00'00''$$

$$(\phi_s) \text{ net spiral deflection to P} = 1°39'59''.6$$

$$\text{difference } (\phi_c - \phi_s) = 3°20'00''.4$$

Lay on TS at $180° + (\phi_c - \phi_s) = 183°20'00''$ for right-hand curves or $176°40'00''$ for left-hand curves.

$$\text{Circle deflection for 10 m chord} = \frac{1}{10}\frac{D}{2} = \frac{D}{20} = 1°00'00''$$

Chord point	Circle deflection	Spiral deflection	Sum
60 (50 + 10)	1° 00′	0° 04′ 00″	1° 04′
70 (50 + 20)	2° 00′	0° 16′ 00″	2° 16′
80 (50 + 30)	3° 00′	0° 36′ 00″	3° 36′
90 (50 + 40)	4° 00′	1° 04′ 00″	5° 04′
100 (50 + 50)	5° 00′	1° 40′ 00″	6° 40′

An alternative solution suitable for angles less than about 7° is shown in Fig. 6.16. For the curve discussed in the previous example using this alternative method would give:

Instrument set up on 50 m chord point and laid on point TS as the back station.

For setting out the 80 m chord point:

$$L_1 = 50 \text{ m}, \; L_2 = 50 \text{ m}, \; L_3 = 80 \text{ m}, \; L_4 = 100 \text{ m}$$

$$\frac{\phi}{3} = 06°40'$$

$$\delta = \frac{(50 + 50 + 80) \times (80 - 50)}{(100 \times 100)} \times 06°40'$$

$$\therefore \delta = 03°36'$$

This value agrees with the value for the 80 m chord point in the previous example.

Above deflection angles of about 7°, minor corrections to these angles need to be made. The tables referred to above contain tables of corrections and explanations of their use.

Control of Excavation

Once the centre line of a road is set out, side width slope pegs (batter pegs) must be positioned to control excavation and filling at the construction stage. Often at this stage accurate cross sections at all 10 m chainages do not exist, and the position of the batter pegs must be found on the ground by a simple trial-and-error process.

To do this proceed as follows:

(1) Study the layout plan, typical cross sections, and any other cross sections there may be.
(2) Compile a schedule of information showing:

- Formation centre line levels
- Differences from above for outer margins
- Centre line peg heights
- Rough distances to top and toe of bank or cut from plan (or sections).

(3) Draw typical cross sections in field book with all dimensions.
(4) Check level to a convenient start point and check centre line peg height.
(5) Set out ranging poles on one side of the centre line to denote the lay of the cross sections, i.e. normal to the centre line.
(6) At first cross section, the assistant takes tape and level staff and goes to the first trial distance, from the centre line to top of cut or toe of bank as the case may be.
(7) The staff is read at this point: if the distance is 10 m from the centre line, the reduced level is 105.30 (Fig. 6.17). This is compared with the ideal level sought, which should be a rise or fall equal to half the distance from the edge of the formation to the top or toe. In this case the formation is 7 m wide, giving a distance of 10 m minus half formation width, over which the calculation for a 1 to 2 batter is applied. In the example shown the desired level is 106.25, which is obviously higher, so that a further position up the slope must be sought to bring agreement nearer.

    With a little practice the method will be found quick, easy and accurate to carry out.
(8) When the desired level is found, a peg should be driven in a further

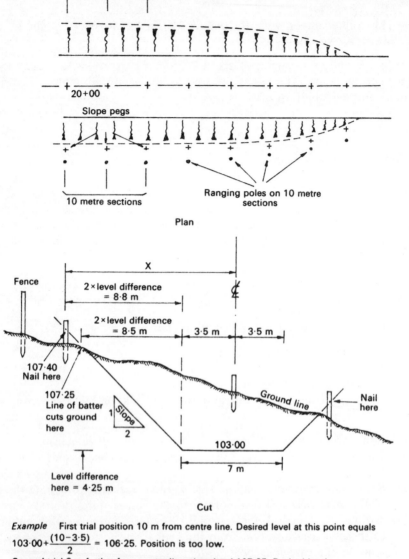

Plan

Cut

Example First trial position 10 m from centre line. Desired level at this point equals
$103\cdot00 + \dfrac{(10-3\cdot5)}{2} = 106\cdot25$. Position is too low.

Second trial 2 m further from centre line gives level 107·25. Desired level
$103 + \dfrac{(12-3\cdot5)}{2} = 107\cdot25$. Peg offset further 300 mm away. Level on peg is +0·15 = 107·40

Fig. 6.17 Roadworks – excavation control – setting out side width slope pegs.

300 mm away from the centre line on cut (nearer on fill), so that the ideal level at the peg is above ground level, and can be marked with a nail.

The assistant should be instructed to move the staff up the peg until the correct reading is obtained, and then to mark the peg level with the foot of the staff.

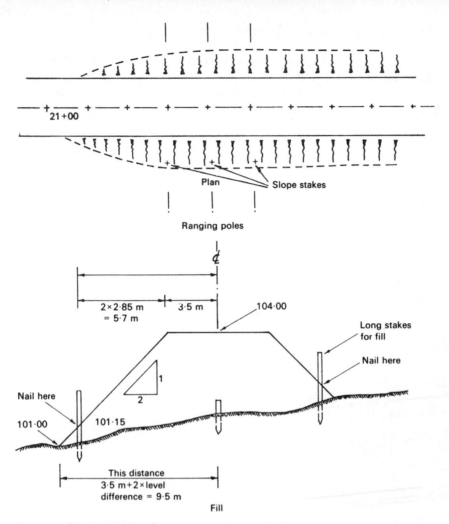

Fig. 6.18 Roadworks − excavation control − setting out side width slope pegs.

Notes
1. Compile schedule of outer margin and centre line levels (pegs)
2. Draw typical cross sections in field book
3. Deduce trial distances for each cross section from contours
4. Mark out sections with ranging poles
5. Check level to area (use peg levels as T.B.Ms)
6. Staff man sets staff at trial distance from centre line
7. Take trial reading and adjust position for level/distance agreement
8. Mark levels on pegs with nails
9. Record details in level book
10. Mark pegs

(9) A nail is driven in at this point to serve as a datum for the batter rule later. The peg is marked with the chainage and the rough figure for cut or fill. If a boundary fence has been erected, record the distance to the peg.

It is useful and time-saving to use long stakes when marking the toe of a bank, since batter rules slanting upwards are normally used with a small

traveller when boning. With a long stake the erection is half complete. A
long stake is also less likely to be obscured by careless tipping.

With a wooden template cut to suit the batter, the assistant and a
labourer can erect the further stakes and position the batter rules at the

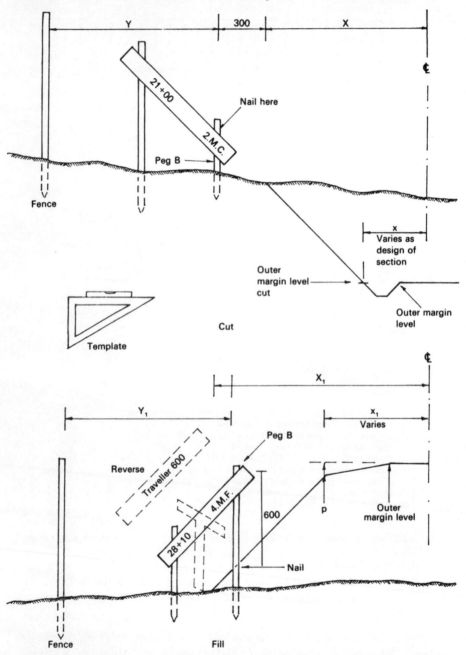

Fig. 6.19 Roadworks – excavation control – setting out batter rules from
existing cross sections.

correct slope with a small boat level. Check for correctness when marking up the batter rules or when doing the check run.

If cross sections have already been taken and plotted, another method can be used which does not involve searching for the correct spot for the batter pegs. From the cross sections, compile a schedule of distances to top of cut and toe of bank from centre line. From inspection of the sections and other data, calculate what the levels should be at these positions.

When setting out the pegs, use the same practice of increasing or decreasing the distance by 300 mm and the level by 0.15 so that fixing batter rules is as simple as before. Discrepancies may be found, in which case the first method can be used, if the field book has been worked up correctly with typical cross section data.

In either method the work should be independently checked and the levelling must always be closed on a BM or TBM of known value and validity. All details should be recorded in the setting out record.

For economy of time and effort it is usually best to work along one side of the centre line for half the day's work, and then back along the other. A good supply of ranging poles is required to work quickly in setting out the sections.

Vertical Curves

Transition from one gradient to another is always achieved with a vertical curve. The form of curve used is usually a simple parabola. Sometimes a cubic parabola is used for economy of earthworks.

The layout and longitudinal sections in Fig. 6.20 show the location and details of levels: allowance for crossfall must be made from centre line levels shown when the profiles are set out.

The degree of detailed information will vary, but operatives still need the right amount of level control to get the work right, i.e. a profile at every 10 m chainage at least. If only the levels for the beginning and end of the vertical curve are available, then the setting out engineer should be able to calculate the rest quickly.

The following notes and figures will assist. The Department of the Environment in the UK lays down certain minimum lengths of vertical curve for different road speeds. On summits this is decided by vision distance. The standard eye height for a seated passenger car driver is 1.08 m and this controls vision distance.

Gradients are expressed as percentages and the grade angle (equivalent to the deviation angle of a horizontal curve) as the algebraic difference between grades (Fig. 6.20). Figure 6.21 shows every possible combination of gradients and values of grade angle. If possible, the distance between the beginning and end of the vertical curve should be an even number of 10 m stations.

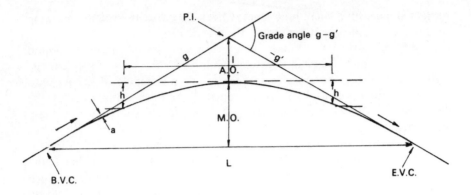

P.I.   Point of intersection of gradients
g.g′   Gradients expressed as percentages
g−g′   Grade angle (algebraic difference)
L   Length of curve
A.O.   Apex ordinate
M.O.   Mid ordinate
a   Offset from tangent
h   Height of passenger car driver's eye (UK 1·08)
l   Vision distance
±   Ascending/descending gradient
B.V.C.   Beginning vertical curve
E.V.C.   End vertical curve
Vertical curves are transition curves, either simple or cubic parabolae. They are classed as
summit or valley curves (sags).

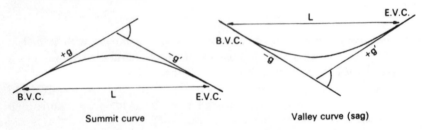

Summit curve   Valley curve (sag)

Fig. 6.20 Roadworks − vertical curves − terms and relationships.

In a parabola the offsets from the tangent vary as the square of the
distance from the tangent point (as the cube in cubic parabolae). The
formulae shown use this to calculate offsets from the tangent grade to
arrive at VC levels.

The procedure is as follows:

(1) Note length of vertical curve (even number of 10 m stations).
(2) Note grades of intersecting tangents.
(3) Note chainage of beginning (BVC) and end (EVC).
(4) Draw sketch to assist calculation.

All possible combinations of gradient are shown with the derivation of their grade angles
(g−g′)

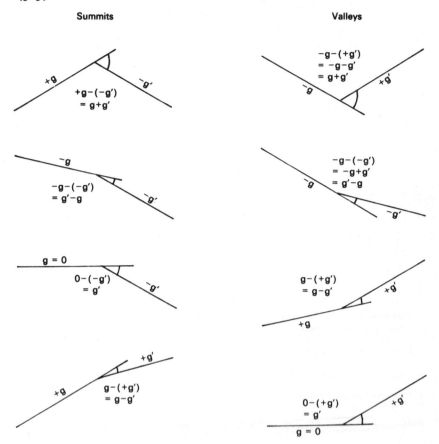

Fig. 6.21 Vertical curves − calculation of grade angle.

(5) Compute (or take off if shown on section) levels of tangent grade for
    every 10 m station.
(6) Prepare a schedule as shown in example.
(7) Calculate a (the offset from the tangent).
(8) Multiply by the square of the distance (in stations) from the BVC for
    successive offsets from the tangent at the other chainage points up to
    the mid-point, i.e. by 1, 4, 9, 16, 25, etc., and insert on schedule,
    both sides of the mid-point.
(9) Subtract (summits) or add (sags) to the levels of the tangent grades to
    arrive at curve levels.

A quick plot of the levels on graph paper will provide a visual check of
the correctness of the work. An arithmetical check is provided by what
are known as second differences, which, if the calculations and additions

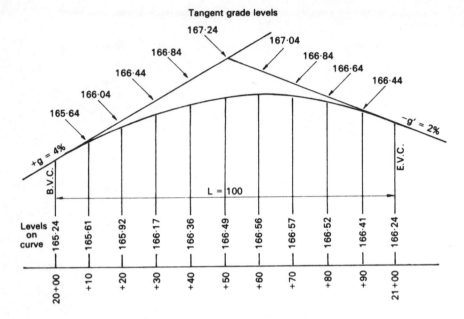

Formulae in relation to vertical curve

1. Rate of level change for first and last point on curve = $a = \dfrac{g-g'}{2n}$
   where n = number of stations

2. Rate of change between all other points = $2a = \dfrac{g-g'}{n}$

3. Apex or mid ordinates = $\dfrac{n(g-g')}{8}$

4. Offset from tangent at any point = al^2 where l is distance from B.V.C. in stations (e.g. at 20+30 = $a \times 3^2$)

5. Distance highest (and lowest) points from B.V.C. (and E.V.C.) for asymmetric curves as below

   Summits from B.V.C. to high point = $X = \dfrac{gL}{g+g'}$

   (ignore sign of gradient)

   Valleys from E.V.C. to low point = $x_1 = \dfrac{g'L}{g+g'}$

   Using the above formulae a setting out schedule of vertical curve levels at 10 metre intervals can be prepared

Fig. 6.22   Vertical curves – calculation of setting out levels.

and subtractions have been done correctly, should be constant. To carry out, extract the differences between curve levels at each station on either side of the highest and lowest points. The second differences between these differences should all be the same.

Level control on the site is now achieved by setting out profiles at a standard offset from the line of the proposed carriageway. The amount will vary on different contracts according to standing instructions; a con-

1. Note the chainage and level of B.V.C. and E.V.C.
2. Take off or compute levels of 10 metre chainage points on tangent grades
3. Calculate rate of level change $a = \dfrac{g-g'}{2n}$
4. Apply a to tangent grade levels with square of distance
5. Check second differences or plot on graph paper

Example

B.V.C. 20+00. 165·24.   E.V.C. 21+00 166·24

$g = +4\%\ g' = -2\%$. $a = \dfrac{4-(-2)}{20} = \dfrac{6}{20}$ = 0·3 per 100 m

= 0·03 per 10 m

L = 100 metres

1	2	3	4	5	6		7
CHAINAGE	STN No.	ORDINATES FROM TANGENTS	TANGENT GRADE LEVELS	LEVELS ON CURVE	DIFFERENCES		REMARKS
					1st	2nd	
20 + 00	0	NIL	NIL	165·24	165·24		B.V.C.
					0·37		
+ 10	1	a	0·03	165·64	165·61	0·06	
					0·31		
+20	2	a×2²	0·12	166·04	165·92	0·06	
					0·25		
+30	3	a×3²	0·27	166·44	166·17	0·06	
					0·19		
+40	4	a×4²	0·48	166·84	166·36	0·06	
					0·13		
+50	5	a×5²	0·75	167·24	166·49	0·06	MID POINT
					0·07		
+60	6		0·48	167·04	166·56	0·06	} HIGH
					0·01		POINT
+70	7		0·27	166·84	166·57	0·06	
					0·05		
+80	8		0·12	166·64	166·52	0·06	
					0·11		
+90	9		0·03	166·44	166·41	0·06	
					0·17		
21 + 00	10	NIL		166·24	166·24		E.V.C.

Note that curve is asymmetric and the high point is displaced from the centre

Using formula 5 from Fig. 6·22, $X = \dfrac{4L}{4+2} = \dfrac{400}{6} = 66\cdot6$

High point is therefore at 20+67. Check√

Fig. 6.23 Vertical curves − preparation of setting out schedule.

venient offset is 450 mm which is near enough for control and far enough away to clear the line of kerb foundations and the like.

It is normal to measure the offset from the front of the proposed kerb line to the nearest face of the profile stake (not the centre). Care should be taken to set the stakes upright and square to the carriageway. They can be positioned first and marked with the appropriate levels when in position. The crossheads must be marked with the length of the traveller in use.

If the design of construction dictates that the kerbs are laid before the pavement slab it is of great assistance to the kerblayers if the stakes are marked with the level of the top of the kerb, so that levels can be transferred with an ordinary 1 m builder's level. A nail driven horizontally makes a good mark.

If kerbs are laid by a slipforming machine it may be necessary to provide level control more often than every 10 m, particularly on vertical curves. The control wire levels should be carefully checked. The method of calculation is no different, only a change in curve station distance is required.

Crossfall Profiles

Another aspect of construction control is the erection of crossfall profiles or padstakes. On all but the simplest of cross section design there are a number of differing crossfalls, e.g. the carriageway or ways, the verge and, on motorways, the hard shoulder.

To get the fall of these various parts of the formation constructed correctly, profiles are erected at all cross-section chainage points and in other places where there are changes of section. They are erected at the bottom of the cut or top of the fill with the heads facing across the carriageway. They are set in pairs to the fall they are to control and should usually be colour-coded to ensure that they are used together and with the correct length of traveller.

On dual carriageway roads there will be at least four different falls per section. Positioning the padstakes on the centre reservation has the advantage of simplicity since there will be only one crosshead on the padstake at a time, so that errors are less likely than if there were four or more permanent ones.

With a long road contract there is no reason why all crossfall profiles should not be of padstake variety, with different lengths of clip-on traveller if required to differentiate between different parts of construction. In any event, there will seldom be less than two crossheads on the outer margin stakes and there should be a different standard traveller length for each crossfall controlled.

Care should be taken that all concerned are aware of the significance of the colours of the crossheads. It is an obvious convention that pads on the central stake are fixed to the side on which they are to be used. If set at standard distances from the centre line they also serve to delineate the edges of the formation.

Standard traveller lengths should be decided upon and a schedule of centre line and outer margin levels at the standard distance from the centre line prepared before attempting to set the levels. If the standard distance is adhered to as much as possible it will be found that fixed relationships between the pads occur frequently, leading to a satisfactory economy in setting level sights.

7 Setting Out Drainage Works

Tasks involved in setting out drainage works may vary from a complete drainage contract to those which are the integral part of almost any other type of contract.

In either case the prime factors involved are:

* Position
* Levels
* Control of excavation.

Position

All drainage runs (except very large diameter pipes) have manholes or inspection chambers at every change of direction or gradient. The positions of these manholes therefore control position, and these points must first be marked by pegs.

The setting out data can be compiled in the usual way by taking off measurements from the setting out drawing and the drainage layout drawing. Checks must be incorporated, and two useful ways of checking the run of drains are:

* A distance check between manholes.
* An orientation check of the line of each straight.

The orientation check can be done by projecting the line of the straights in either direction on the plan and noting where these projections intersect other parts of the site layout.

It is important that the MH pegs should be marked distinctly and correctly, as foul and surface water systems frequently run parallel and in close proximity. They should be referenced at once if excavation is to follow.

If trenches are to be dug by machine then it is generally preferable to set up offset profiles clear of the trench line. Consult with the general foreman about spoil disposal and set profiles on the opposite side to the

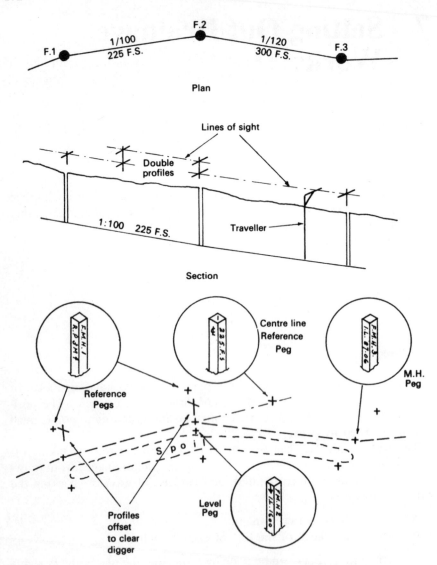

Fig. 7.1 Drainage marks.

dumping area. They should be clear of the centre line by half the machine width plus 300–600 mm. The traveller must be L-shaped.

With hand excavation it is generally better to set up striding sight rails. They can be either immediately over the MH pegs or set back along the line of the trench by a metre or so to allow the MH to be dug. Allowance should be made for the different level of the false invert at these points when fixing sight rails.

Levels

In the pipelaying stage it is often more helpful and productive to fix a level peg near to each manhole so that operatives can transfer the invert level to the trench bottom easily, without continual reference to the sight rail and traveller. These pegs may also serve a useful purpose in setting the cover levels of the manholes when built. They should be marked in dimensional terms (as described earlier), rather than with a level value, e.g. IL Down (or Up) 950.

Remember to allow for bedding dimensions when deducing traveller lengths, which must also allow for pipe thickness.

When setting levels, or level marks relating to levels, the following simple practice will ensure that the level set bears a round figure relationship with the construction level. Subtract the construction level value from the height of instrument and note the last three figures. Signal the staff holder to move the staff up or down against the peg or stake to be marked until these last figures are read irrespective of the whole metre. The staff holder now marks the peg level with the foot of the staff and the complete reading is booked. The reduced level will now bear a simple relationship with the desired level, and measurement for operative work is less liable to error.

On occasions, in large buildings with deep concrete raft foundations, drainage pipes of cast iron are placed before the raft is cast. When this is done the levels of the tops of all connections cast in with the concrete must bear the correct relationship to the finished floor level. Level marks should be provided for the plumbers and the level of the finished pipework should be checked before the concrete is poured.

Control of Excavation

Concrete works in drainage schemes in general require similar marks and control to those of buildings and deep foundations, and the practice for these can be safely followed.

When working up the information for setting out, the specification as well as the drawings for details of pipe bedding and surround should be consulted. These dimensions will affect the lengths of any travellers used with the sight rails for excavation control.

(1) Study plan, sections, manhole schedule and specification for details.
(2) Work up field book with all dimensions and diagrams for setting out.
(3) Set out MH positions.
(4) Set in offset pegs and any necessary centre line reference pegs or otherwise to re-locate MHs after digging. Record details in field book.

(5) Set up profile stakes to clear line of digger and opposite to proposed spoil dump (consult foreman about this).

(6) Check level to convenient position for setting sight rails on profile stakes.

(7) Set sight rails at correct levels, mark length of traveller on back.

(8) Mark all pegs clearly.

(9) Check all work. Record details in setting out book.

(10) Make sure that operatives understand your marks (make a sketch for the foreman in a duplicate book).

8 Setting Out Building Works

The size of the task may vary enormously but the principles set out in Chapter 4 remain the same. Buildings must be controlled during construction for:

- Position
- Size and shape
- Level
- Verticality.

Position, size and shape are originally determined and marked in terms of corner pegs representing the outside faces, which must be superseded at once by reference pegs or profiles clear of the work area.

Level is determined originally by establishing a datum on site and then by transferring levels to profiles from which construction work can be carried on without constantly requiring the use of an instrument by the engineer.

Verticality can be controlled by the use of plumb lines on low rise structures, but there is no substitute for an Autoplumb on high rise buildings.

Initial Marks

The method used to position initial marks will depend on several factors:

- Accuracy required
- Number and layout of buildings
- Planned control and setting out data provided.

On a small site the setting out drawing will often provide all the dimensional data needed to locate the corners from existing ground detail or original survey marks. Even if it does not, there is no difficulty in taking off trial dimensions and reconciling them with distances measured on site.

On a large site one of two conditions is likely to exist: either there is a planning or setting out grid, or the layout is such that it will be economical of time and more accurate to superimpose setting out base lines on the plan. They may or may not cross at right angles, according to the layout. In either case, the first priority is to establish marks which relate the grid or base lines to the site. This has the added advantage that in doing so any discrepancy between site plan measurement and the ground will be revealed and can be allowed for.

With a planning grid, the setting out detail must be worked up as follows:

(1) Transfer to field book any dimensions given for the location of key grid lines.
(2) Take off (by scaling if no other detail available) check dimensions to permanent detail which will show the correctness of the position and orientation of the key lines when set out.

If proper planning consideration has been given to the key lines they should be located where clear lines of sight will persist for longest as construction goes on. If not, however, the engineer should examine the layout carefully to see which lines are likely to meet this condition best and at the same time provide the maximum future economy of measurement. The lines chosen should then be set out first; there will be no great difficulty in relating the original dimensions to them in the case of a rectangular grid.

The key point in this method will be the intersection of the chosen key lines, and the data for fixing this should be carefully examined to see which promises the best result for fix (i.e. absolute position) and for orientation. The appropriate data should be used in that order for establishing the key point.

Having established the position of the key point and the orientation of one of the lines running through it, the right angle relationship of the grid lines should be established as follows:

(1) With a theodolite set up over the key point, sight on the furthest point visible on the line which has been oriented. The plate should be set at a small angle rather than zero.
(2) Read and book this small angle in the field book.
(3) Unclamp the top plate and turn either way through approximately 90°, reclamp.
(4) Using the top plate slow motion screw and viewing the top plate reading, bring the reading to a figure which is exactly 90° or 270° different from the original angle. Care should be taken not to disturb micrometer settings during this operation.
(5) Line in a distant peg held by an assistant on this line; a temporary fine mark should be made by the assistant at this stage.
(6) Change face on the instrument and now observe the angle set out on both faces.

(7) In the unlikely event of the mean angle being 90°, proceed to the erection of permanent sighting marks outside the area of construction.

(8) If, however, the mean angle differs from 90°, direct the assistant to make a further fine mark at a reading on the original face which will produce a mean of that figure. Re-observe on both faces until corrections required produce the mean.

Though this procedure may seem tedious it will repay the care taken. As all experienced engineers know, accurate right angles are difficult to set out. The engineer must decide what tolerance to allow by simple consideration of likely error.

Note: 1 minute $\simeq$ 100 mm at 350 m
6 seconds $\simeq$ 10 mm at 350 m

The next requirement is to give permanence to the fix and orientation by constructing marks which will not be disturbed as the work on site proceeds. This is best done by siting marks outside the scope of the new work and referencing key points so that, if disturbed, they can be recovered without the necessity of going through the whole initial process again.

The ends of the key lines should be marked by pegs concreted in position at suitable places on the perimeter of the site. They will not necessarily be at grid intersections so the measurement of their position should be done with care. If the site is a large one it may be necessary to use correctly tensioned taping and correcting for temperature or EDM, if available.

As setting out proceeds it will constantly be necessary to align marks in the line of the grid and it is a good plan to set up sighting targets over the permanent pegs. This can be done by erecting a white board supported on a framework over, or slightly back from the permanent peg, and then, by instrument sighting, marking a black vertical line on it exactly in the grid line. The line should not be too thin or it will not be visible even through an instrument telescope from normal site distances.

The key point from which the grid was originated should be temporarily referenced and made permanent by concreting. A guard rail should be erected around it at once. If the key point is in such a position as to make it inconvenient to have an obstruction to free movement, it is advisable to sink it below the normal ground level by 100 mm or so, fill the pit above it with sand and cover with a steel plate. Site traffic can then pass over without fear of disturbing it, and the mark can easily be uncovered on those occasions when it is essential to re-use it.

If the site plan has no grid and the site is such that it will be economic to use this method for setting out there is a simple way to superimpose one on the layout.

Prepare a suitable rectangular grid on transparent material (preferably a stable one such as 'Permatrace' or the like; otherwise use tracing paper). Check it carefully for squareness and dimensional accuracy, then move it about over the site layout so that the most convenient position and orientation can be found. Prick through at least four control points and reproduce the lines carefully in pencil on the plan. Extend the lines until they cut original traverse lines or pass close to important fixed detail. Extract the data of these intersections or positions from the plan and set out as described.

It is not vital that base lines should cross at right angles. It may be more convenient to have them parallel to the orientation of building faces of major groups of buildings. Their angular relationship and points of intersection, however, must be known and checked against the ground.

It is always advisable to check the separation of buildings from the detailed drawings and the specification and carry out total dimensional checks against the scaled plan positions and the known relationships of the base or grid lines with the ground. If building locations are given in terms of coordinates the task is somewhat simpler, since once the grid has been fitted to the site and checks done, all positions in terms of the grid should fit.

Once the main axes of the grid are established, other points can be made permanent so that all subsequent detailed setting out of buildings becomes a more local affair of offsetting from the nearest convenient grid line without the necessity of measuring long distances each time. The principle of independent dimensional checks must still be adhered to in order to guard against mistakes.

A gridded plan is much more satisfactory to use as it is possible to extract the positional data of buildings in terms of coordinates by scaling. Errors likely to arise from distortion of the plan can be detected at once by the disagreement of the grid lines with the scale; allowance can be made at the time by taking the mean coordinate values with the certainty that if the coordinate differences match the building dimensions, and the grid has been reconciled with the ground, all will be well.

Checks can also be made more easily since, once coordinates are in use, linear and angular relationships between parts of the layout can be calculated with ease and accuracy. On sites flanking roads where there is already a building line in existence, or when one is shown on the site plan, it is most important to make sure that this is marked first so that there is no encroachment with subsequent setting out.

In the main, building lines are set back a fixed distance from road centre lines, which are likely to be more regular than kerb lines. On an existing road the centre line should be found by measuring the carriageway at intervals and making marks at the centre points (a pipe nail driven into the wearing course makes a good mark). A visual check for straightness or regular curvature is often sufficient to ensure that the line so produced is an accurate starting base for setting in the building line. At doubtful points, several check measurements to other detail should be

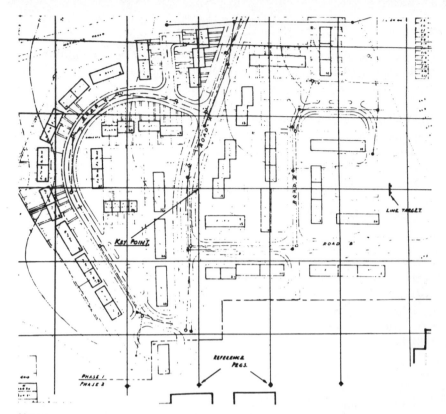

Notes

1. The grid may be pre-contract planned or superimposed by the engineer
2. Grid lines should be referenced at convenient positions near the site boundaries and clear of construction work
3. The right angle relationship of the grid lines should be established at the key point
4. Line targets are useful on main or important much-used lines

Fig. 8.1 Coordinate setting out.

made. It should always be remembered that any setting out measurement is only as accurate as the data on which it is based.

Preparing Data

Having established base lines or grids, or other starting points from which to measure, it is essential to develop a systematic approach to the assembly of data for use.

All setting out drawings should be checked for accuracy of dimensions; running dimensions should be totalled and detailed dimensions checked. Discrepancies may well come to light at this stage and should be noted and checked with the designer responsible for the layout. Alterations should be confirmed in writing by the designer. Where alterations are made they should be clearly annotated with the date and authority for alteration.

It is a good plan to break the main setting out drawing into parts which are to be set out in a practical sequence, and to transfer the data to sections of the field book so that each finite piece of setting out is complete in itself, i.e. it includes all detail necessary for both setting out and checks. Liberal use of sketches should be made so that the work can easily be checked independently.

Sequence of Operation

(1) Set out corner pegs from data.
(2) Check diagonals (for large buildings this may be done instrumentally).
(3) Compare position with check dimensions.
(4) Set up stakes for profiles or show assistant where this is to be done. Specify offset.
(5) Check level from nearest TBM for SFL.
(6) Set profile boards in relation to SFL.
(7) Using string lines, transfer line of corner pegs to profiles and mark.
(8) Mark relationship profile height to SFL on at least two profiles. Show traveller length for foundation excavation.
(9) Mark foundation limits on profiles.
(10) Recover corner pegs for re-use after checking.

The above procedure applies mainly to buildings with normal strip foundations. Where large, deep excavation pits are to be dug, the limits of the excavation should be marked by any convenient means (pegs) and temporary profiles provided for control of depth, leaving the main setting out profile around the perimeter to be fixed when machine work is finished. This will avoid the danger of disturbance during the major earth-moving operations and will not delay the setting in of details of position of lift shafts, foundations, tower bases, etc.

The setting out marks which control the plan position of the building as a whole will have been set back well outside the immediate area of excavation and concreted in accordance with the detail elsewhere. From these marks, the limits of various parts of the building can be marked on a continuous profile around the edge of the finished excavation and finite points set by the use of plumb lines from stringlines stretched tautly between corresponding marks.

The profile should be set, as before, at a fixed relationship to structural floor level.

On the sites of buildings with large floor areas and of low rise (e.g. factories) which are likely to have a column support structure (either steel or concrete), it will be necessary to set in reference pegs from which the position of the column bases can be marked with considerable accuracy when the floor slab is cast. It is important to choose the right time to do this; if done too early, the pegs, which must be set to fine limits, will be disturbed during the erection of formwork and concreting. It is best to consult with the General Foreman, for it will often have to be done to fit very closely with the construction work timetable. The reference should be transferred to small iron plates set in the floor slab while it is still green, and centre punch marks made in the plates when they are firm.

Vertical Control of Tall Structures

The need to plumb upwards before construction and the relative inadequacy of plumb lines, with the associated difficulty of plumbing accurately over a mark with any but short lines, has led to the development of a number of optical plumbing devices. They are special-purpose instruments but their use and cost is more than justified in terms of time-saving and proper control of verticality in tall structures (e.g. high rise building, cooling towers, etc.).

Some of these instruments depend solely on sensitive bubbles for level and turn the line of sight into a vertical plane with either a fixed or adjustable prism.

The Autoplumb

The Autoplumb (Fig. 8.2), in addition to bubbles, has a compensator similar to that of an automatic level. This ensures a much higher degree of accuracy in the vertical line of sight and incorporates a tilting mechanism for the deflecting prism which is useful both for measurement of any departure from the vertical and in enabling the engineer to keep a check on the adjustment.

The instrument is basically an upward sighting telescope with a compensating prism in the optical system to ensure against minor dislevelment (up to ±15 minutes of arc), and gives a truly vertical line of sight ±1 second of arc. It is fitted on a traverse base with a sliding head and is interchangeable with traverse targets or a traverse theodolite by the same maker. The downward sighting telescope is for centring over a marker; it is not compensated, but depends for accuracy on the levelling of the base as do other optical plummets.

The line of sight is reflected from the horizontal by a pentagonal prism. This is controlled by a micrometer drum on the top of the instrument. One revolution of the drum tilts the line of sight through an angle of 1 mil (tan 0.001), giving direct measurement of deflection from the vertical, i.e.

$$\frac{\text{Drum Revs} \times \text{Height}}{1000} = \text{Displacement}$$

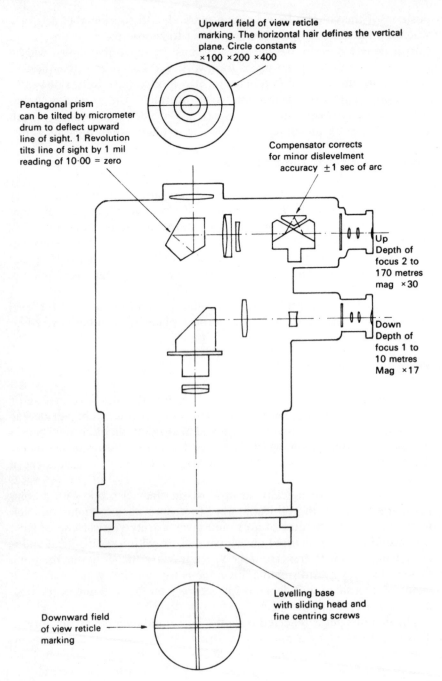

Upward field of view reticle
marking. The horizontal hair defines the vertical
plane. Circle constants
×100 ×200 ×400

Pentagonal prism
can be tilted by micrometer
drum to deflect upward
line of sight. 1 Revolution
tilts line of sight by 1 mil
reading of 10·00 = zero

Compensator corrects
for minor dislevelment
accuracy ±1 sec of arc

Up
Depth of
focus 2 to
170 metres
mag ×30

Down
Depth of
focus 1 to
10 metres
Mag ×17

Levelling base
with sliding head and
fine centring screws

Downward field
of view reticle
marking

Fig. 8.2 Vertical control − the Autoplumb optical system (Hilger & Watts).

The micrometer drum is divided into a hundred divisions working over a thimble (as for the gradient drum on a tilting level), numbered from 8 to 12. The thimble scale records whole revolutions and the drum scale 100th parts of a revolution (0.01 mil). When the instrument is level and in adjustment, a reading of 10.00 represents a truly vertical line of sight.

To test the adjustment of the instrument, set up over a mark and sight on a target vertically above. (The target can be adjusted by an assistant or can be an offset scale.) When a mark is suitably near the horizontal hair in the field of view, turn the micrometer drum so as to intersect it exactly, and note the reading R_1. Turn the instrument through 180° (there are stops on the base to ensure this) and repeat, giving a reading R_2.

Then $\dfrac{R_1-R_2}{2}$ = angular displacement of mark from the vertical. If the instrument is in adjustment, $\dfrac{R_1+R_2}{2}$ will equal 10.00. If there is an error in adjustment, then a drum reading of $\dfrac{R_1+R_2}{2}$ is the correct setting for a vertical line of sight. The field of view upwards shows an object as seen if looking vertically upwards naturally. The horizontal line of the reticule defines a vertical plane at 90° to the horizontal. Displacement of a target in a direction along the axis of the telescope defines departure from this vertical plane and can be observed and measured. Turning the instrument through 90° defines a vertical plane at right angles to the first, and the intersection of the two a point immediately above the ground mark.

An ordinary theodolite can be used in a similar way with a right angle eyepiece attachment, but without the benefit of the compensating device it would be necessary to define the vertical as a mean of four marks made in planes 90° apart.

If the test of adjustment, previously described, has been done and the mean of $R_1 + R_2$ correctly set, then the line of sight will be vertical and there is no need to sight in four planes with an Autoplumb unless the assistant finds difficulty in aligning the target with the centre of the plane of reference.

The instrument can also be used to plumb a point directly beneath an overhead mark (Fig. 8.4).

Using the Autoplumb in Multi-storey Construction

In multi-storey building construction, plumbing up by ordinary plumb-lines for the formwork may keep each level reasonably vertical, but control of squareness is much more difficult, and the use of the Autoplumb will more than repay the cost of the instrument.

The original plan position of a building will have been defined by reference pegs and the subsequent perimeter profile. As soon as the ground slab is poured, a grid should be established inside the confines of the building to transfer measurement from the original setting out. The marks will define lines at right angles to the building faces or the column structures. They should be positioned so as to provide vertical lines of sight through service (or purpose-made) apertures in succeeding floor slabs. At least three such marks, preferably four, are required for simple plan layouts. For odd shapes of floor plan, more may be required.

To transfer the grid from floor to floor so that the building and column structure remains square and plumb, the Autoplumb should be used as follows.

Site the Autoplumb over each mark in turn and, with the instrument roughly orientated to one face of the building, direct an assistant by voice or signals to move a straight edge on a transparent panel to the upper floor slab, until it is coincident with the hair line of the reticule. Turn through 90° (to the next automatic locking position on the head of the instrument) and repeat the process. The intersection of the two planes marked by the assistant denotes the vertical grid mark which should be checked at once by observation in the other two planes. A wooden frame with a perspex sheet to fit into the apertures is a suitable device for a marking panel. China or glass-marking pencils can be used to mark it. The grid established should be immediately offset to points on the floor slab itself by direct measurement between any two of the upper marks. When this is done the targets should be removed and the apertures covered with safety plates.

The offset positions should be so sited as to run outside the lines of any column structure involved, so that they can conveniently be used for setting out formwork. It may not be necessary to carry out this procedure for every floor, though there is much to recommend doing so. Where the upper levels are reached, voice communication will be difficult, if not impossible. A radio or field telephone should be used. As a last resort, a black signal board should be used to convey orders to the assistant. Binoculars can be used to assist in reading from heights: Fig. 8.3 should make the process clear. Properly organised, the transfer of the grid for each stage can be very quickly done; as soon as the grid is marked on each floor, much internal setting out is eased.

On tall buildings with open floor plans and curtain wall construction, where the column structure is visible from the outside, the same result can be achieved with an ordinary theodolite, but with the need for rather more elaborate arrangements. Provided a diagonal eye piece is available for the theodolite, marks can be made at the intersection of planes at right angles to each other. These planes are defined by marks outside the confines of the building and sufficiently distant to keep the vertical angle of sight below 70° or so for the topmost marks.

A 20-second theodolite plate bubble is usual (i.e. the level plane defined is accurate to within 20 seconds of arc), and this is not comparable with the Autoplumb and the compensator. Nonetheless, if the trunnion axis adjustment is correct, the instrument is carefully levelled and the marks are the mean of right and left face observations, satisfactory results can be obtained.

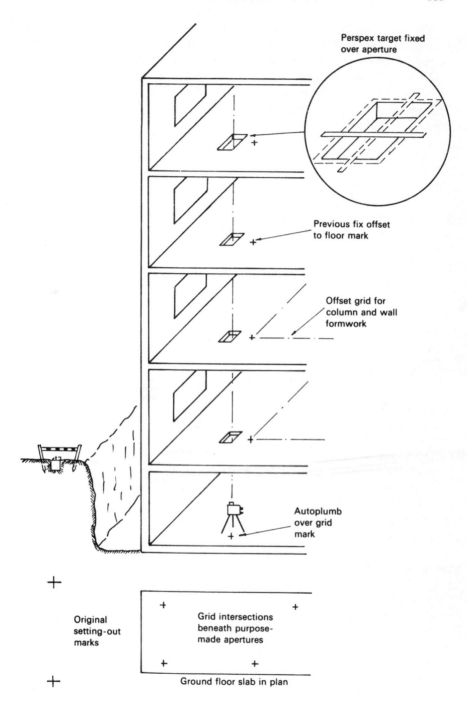

Perspex target fixed over aperture

Previous fix offset to floor mark

Offset grid for column and wall formwork

Autoplumb over grid mark

Original setting-out marks

Grid intersections beneath purpose-made apertures

Ground floor slab in plan

Fig. 8.3 Vertical control of tall buildings.

The exterior grid, which needs more setting up positions, is more elaborate and time-consuming to set up and the marks themselves are more prone to disturbance than internal grid intersections marked on floor plates. On confined sites, it may also not be possible to set up far enough away from the building to use this method. For cooling towers, chimneys, etc., the Autoplumb is a very necessary piece of equipment.

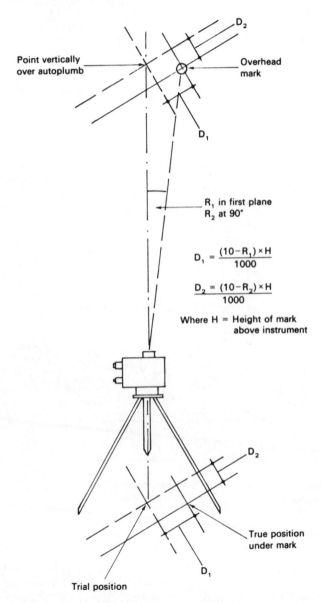

Point vertically over autoplumb

Overhead mark

D_2

D_1

R_1 in first plane
R_2 at 90°

$$D_1 = \frac{(10 - R_1) \times H}{1000}$$

$$D_2 = \frac{(10 - R_2) \times H}{1000}$$

Where H = Height of mark above instrument

D_2

True position under mark

D_1

Trial position

Fig. 8.4 Vertical control − plumbing from a mark.

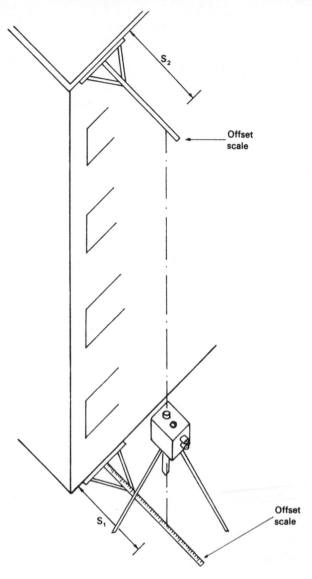

Fig. 8.5 Vertical control – checking verticality.

Lasers

A laser mounted to point vertically up or down could be used as an alternative to the Autoplumb. This has the advantage that the desired vertical line is visible and so only one person is required. Once set up the engineer could ascend to each floor in turn to mark out the building line. A transparent or frosted screen could be used to image the laser spot. When the last point has been marked the engineer should return to the laser to check that it has not been disturbed. However, many builder's lasers would not be able to give the same precision as the Autoplumb but they may still give an acceptable result.

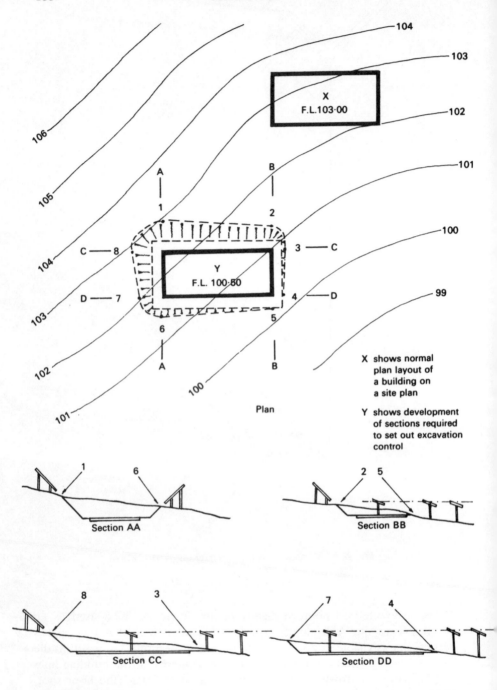

Fig. 8.6 Position of batter rules and profiles for excavation control on a sloping site.

Plan

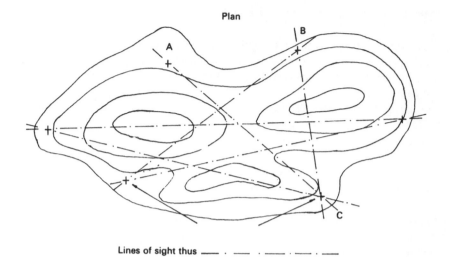

Lines of sight thus ——— · —— · ·—— · —— · ——

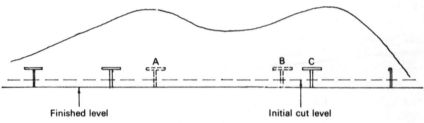

Finished level   Initial cut level

Section through site to be levelled

Example
  Levels for profiles
Finished level   105·26
Initial cut above   0·10
                          105·36
Traveller   1·25
Profile height   106·61

Required staff readings
(instrument height 106·82)
For profiles 1·56
For mark on stake to suit
1.25 m traveller 8.21

Free standing
traveller for
use of banksman
or machine
operator

Fig. 8.7   Site levelling control.

Site Levelling and Control of Excavation

On sloping sites, where it is usual for buildings to sit on level shelves cut into the slope, it will often be necessary to cut such positions before the normal profiles for control of foundation excavation and construction can be set. To control such excavation, the limits must first be marked, and profile and batter rule control marks erected for the excavator operators.

Unless the slope of the site is very steep, the setting out drawing may not show the limits of excavation and the new banks; in this case, careful study of the plan for each building is necessary and cross sections should be constructed, using graph paper. The lines of the faces of the building are extended and a note made of where these intersect the contour lines.

The dimensions of the level surround should be allowed for, and the specified slopes from its limits drawn in to cut the line of existing ground on the cross sections, the distances from the building where these points occur being noted. If a grid is in use, it may be possible to mark their position on the ground by reference to this; otherwise it is necessary to stake out the corners of the building first and relate the excavation limits to them.

The excavation is now defined in area for the excavators, who will additionally need some means of level control. On the uphill side this can be provided in the same way as on road works, by the provision of batter rules, marked with the depth of cut and set to the chosen batter.

For the final level of the sites, profiles should be set up on the downhill limits and beyond them longer stakes with crossheads at the same level. The line of sight across the tops defines the finished cut level when used with a traveller of correct size. A free-standing one should be made up for use by the banksman.

When the whole of a rough or undulating area is to be levelled to a certain level, the limits are sought out by use of a level and staff, as though contouring. Profile stakes are erected on the line (or contour) which defines the finished level, and crossheads fixed for use with a free-standing traveller. The stakes should be so distributed that lines of sight across them in various combinations cover most of the area to be levelled. Earth-moving machine operators can work from these marks without danger of under- or over-cutting. Frequent checks on the work should be made when near to finished level.

9 Setting Out Tunnelling Work

Two main types of setting out operation arise in tunnelling work:

* Straightforward headings with access at ground level.
* Headings where initial access is at the bottom of a shaft.

For obvious reasons, tunnels are usually driven both ways at once. Sometimes intermediate shafts are sunk to increase the number of working faces, or for some other engineering reason.

The accuracy required in the surface survey varies directly with the length of the tunnel. For example, in a 3500 m tunnel an error of 1 minute in line will result in a misclose of 1 m between faces, apart from any positional or level errors. Even over a distance one tenth as long, it is obvious that care is needed.

Depending on the ground and the length of the tunnel, the surface survey may be a traverse, or a triangulation or trilateration, or a combination of all three.

Survey stations should be sited carefully to give the best possible conditions for angular and linear measurement. They should be concreted in and referenced before observations are taken. Angles must be meticulously observed on both faces and often enough to ensure a very small total misclose.

Measurement of triangle sides or traverse legs will depend on the equipment available. Distance measuring equipment is the most convenient, but may not always be available, so when taping is done it should be carefully carried out with checked tapes, supported and correctly tensioned. Distances must be corrected for slope (and for temperature if the ambient conditions are very different from the normal 20° C). Slope corrections are best derived from the properly levelled heights of stations and points of change, and these must be obtained by meticulous check levelling.

The survey should be computed in rectangular coordinates, and the ends of the tunnel centre line and any changes in direction expressed in coordinates on the same grid.

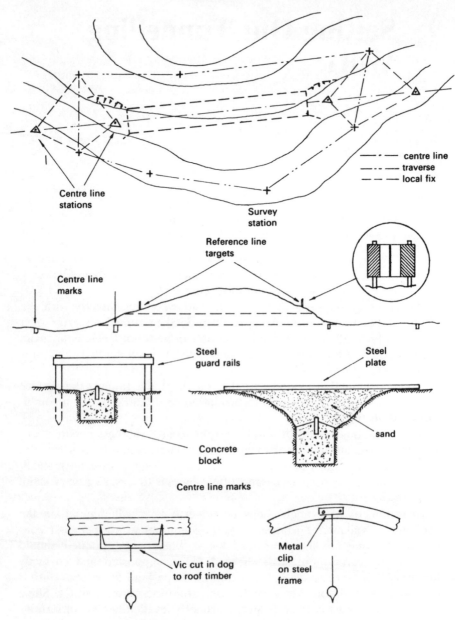

Fig. 9.1 Tunnelling – marks and survey.

Headings with Access at Ground Level

For headings with access at ground level, proceed as follows:

(1) Select two positions on the centre line back from the proposed tunnel

mouth at each end, on which to set up permanent reference marks for line and position.

(2) From their coordinate values compute the bearing and distance from the nearest surface survey stations. Set out and mark with pegs initially.

(3) Check the computed bearing of the tunnel line with the observed bearing between the two marks, using more than one surface station as a reference object when set up over one of them.

(4) Repeat any necessary operations if any error disclosed falls outside the required tolerance.

(5) Erect simple sighting marks for the initial opening of the heading.

(6) Transfer the line to marks on the roof as the heading progresses.

(7) Provide level control by normal levelling procedures from a local TBM. Special short levelling staves are required in small headings, and continuous checks are necessary.

(8) Make the reference marks permanent, and protect against damage.

Where instruments are to be used inside the heading, proper arrangements must be made for instrument lighting and booking, as well as for the illumination of staves and marks. Some form of opaque screen is necessary to illuminate a suspended plumb line.

Nearly all modern theodolites are equipped with internal lighting systems for the reading scales and diaphragm. They are battery powered by clip-on battery packs and simple plug connections. The diaphragm of a level may be illuminated by reflected light from a light surface held near the object glass and lighted by a torch or lamp.

Headings with Access from Shaft

When a tunnel is to start from the bottom of a shaft the procedure is very similar to that just described. One of the positions on the centre line should be the centre of the shaft, which can be marked out so that excavation can start.

When the line is checked and passed, a further mark is required in the same line, and preferably in the direction of the heading, to serve as a reference object later. The first and third marks should be carefully protected against the time when it becomes necessary to transfer the line to the heading at the bottom of the shaft. The line should also be transferred to shaft-head marks as a basis for any shaft works construction. A level datum is similarly required.

When the operating level for the tunnel is reached, a further datum can be established at the bottom of the shaft by direct measurement with a suspended tape under correct weight tension and hanging freely vertical.

The normal way of transferring the line to the bottom of the shaft is by suspending two piano-wire pendulums in the line of the heading right down the shaft. Their motion is damped by using heavy weights in sludge buckets, and the line so delineated is transferred to the heading by the methods described in the notes following (see Fig. 9.2). With the advent of laser beams it may be possible to supersede this method; certainly the laser beam is a most convenient way of maintaining the heading line, once it has been established. The verticality of any beam and its accurate positioning on line at the top of the shaft, the conditions of refraction in the shaft and the spread of the beam all pose minor problems.

If the shaft is deep and there are heavy-duty ventilating fans producing strong air currents, these must be turned off for some 12 hours or so to allow the pendulums to settle down.

For small headings, elaborate arrangements may not be necessary. They are common in drainage and cable duct work in built-up areas, are usually hand-driven and run between shafts which may be intervisible, or it is possible to run a line over the surface between the ends.

Marking the line across the top of the shaft and dropping simple plumb lines down from this will define the line sufficiently accurately for the tunnel men to drive the heading to line over the short distances common to such works.

The level datum is easily transferred from the surface, shafts are usually not deep, and level is maintained by the miners themselves. Small building lasers currently available certainly make this type of operation more easy to carry out.

Transferring the Line

When the pendulums have settled down in their swing, they will never be quite still despite the heavy weights and damping; a theodolite must be used to transfer the line to marks in the roof of the heading.

There are two ways of doing this, one by lining up the instrument accurately with the line of the plumb lines and the other by using the Weisbach triangle method. Neither is particularly easy to do, though the first method is made much easier if a special type of sliding head is available for the theodolite. Some traverse theodolites are provided with this arrangement, which permits fine adjustment of the sliding head in two planes at right angles by the use of slow-motion screws.

If using this method, proceed as follows:

(1) Position the instrument in the heading, lined in as accurately as possible with the line of the plumbs.
(2) Level carefully and focus on each wire in turn without, if possible, using the plate slow motion between sightings. Now measure the angle, if any, between them and see from its value to which side of the centre line the instrument is displaced.

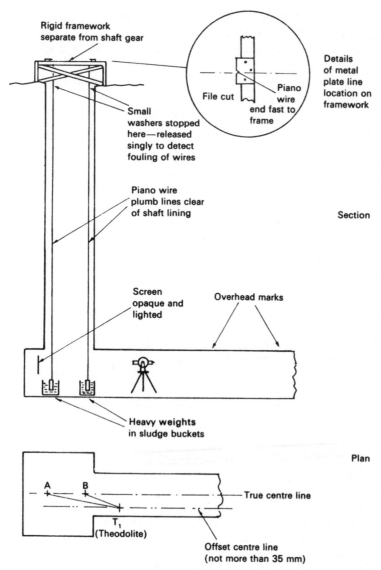

Rigid framework
separate from shaft gear

Details
of metal
plate line
location on
framework

File cut

Piano
wire
end fast to
frame

Small
washers stopped
here—released
singly to detect
fouling of wires

Piano wire
plumb lines clear
of shaft lining

Section

Screen
opaque and
lighted

Overhead marks

Heavy weights
in sludge buckets

Plan

A   B

True centre line

T₁
(Theodolite)

Offset centre line
(not more than 35 mm)

Fig. 9.2 Tunnelling – line transfer – Weisbach method.

(3) By using the distant wire as the RO an angle reading of a few seconds will indicate displacement to the left of centre, while a reading a few seconds short of 360° indicates displacement to the right.

(4) Very gentle movements of the sliding head in the required direction can now be made until the instrument is exactly in line, as indicated by the fact that a mere change of main focus brings each plumb line in turn into coincidence with the vertical hair line of the reticule.

It is quite difficult to detect which plumb line is being viewed, as at these distances change of focus makes one disappear completely. It is a good plan to clip a slip of card to one of them in the field of view to make it easy to recognise.

When the instrument is satisfactorily in line it should be laid carefully on the far wire, then transitted, and a mark made on a roof timber in the heading. This operation should be repeated at face right, and the mean position of the two marks then compared by laying on the near wire and repeating the process.

The mean mark should be adopted as the centre line. Two such marks as far apart as possible should be made. As the heading advances, the theodolite can be plumbed under successive marks and the line extended. It is useful to have an instrument with a central acorn mounted on the telescope at the trunnion axis so that centring under rather than over a mark is made easier.

Weisbach Method

With the Weisbach method no attempt is made to set up in line with the plumb lines, instead the theodolite is displaced slightly to one side (not more than 35 mm). The small angle at the instrument subtended by the wires is read many times on both faces to achieve as accurate a result as possible. The other angle at the far wire is then computed and used to set a line parallel to the true centre line. The offset is then calculated. Either the true line or the offset line can be carried through as described.

Calculation of angle BAT (Fig. 9.2) is simple. Using the sin rule to solve triangle ABT_1,

$$\frac{AB}{Sin\ T_1} = \frac{BT_1}{Sin\ A}$$

$$\therefore\ Sin\ A = \frac{BT_1.\ Sin\ T_1}{AB}$$

Conditions in tunnels are often very difficult and the work is exacting. The engineer must be most careful when observing and must constantly check his or her work. It is essential to have properly trained assistants who understand the accuracy and sighting requirements and are capable of measuring and marking accurately. Setting out work must often be done between shifts.

Level datums must also be carried through, and it is quite sound and usual to do this on the roof of the heading. The engineer must get used to the different booking and reading practice when the staff is used upside down against the roof.

10　Setting Out Marine Works

The main problem likely to arise in setting out marine works is obviously the location of construction works, such as piling, beyond the low tide positions. It is sound practice to use a grid overlay to the layout plan, so that offshore positions can be expressed in terms of coordinates and their location fixed by the intersections of sight lines, which are easy to compute when coordinates are in use.

A common problem is setting the position of piles which are being driven from a floating platform. A convenient way of doing this is by the erection of leading marks on shore, akin to normal navigational practice. The main line of the piles is denoted by one pair of marks, and the position of individual piles by a second pair for each pile, or by a line given by the engineer with his theodolite, intersecting the main line (Fig. 10.1). Alternatively, an EDM with tracking facility mounted on a theodolite, or electronic tacheometer, could be used measuring to a prism on the floating platform.

Communications are vital in this type of operation but in the days of portable VHF radio sets this should not be a problem.

The height of the marks must be carefully considered to ensure that they are visible at all conditions of tide. They must also be erected carefully and checked for plumb in the line they mark. It is possible to arrange the near mark of a subsidiary set to move on an arc to denote the line of intersection of successive piles.

When siting marks inside coffer dams it is important that the trunnion axis adjustment of the theodolite is correct and that the plate is levelled carefully. It is sometimes possible to use optical plummets, but there is always the difficulty of finding an instrument platform free from vibration or the possibility of movement in position.

Intersection sightings involving transitting the telescope should *always* be done on both faces.

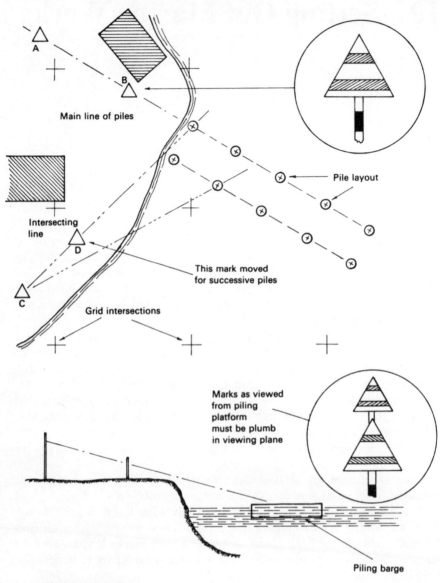

Main line of piles

Pile layout

Intersecting line

This mark moved for successive piles

Grid intersections

Marks as viewed from piling platform must be plumb in viewing plane

Piling barge

Far and near marks must be high enough to be seen at high and low tide times

Fig. 10.1 Marine works – piling marks.

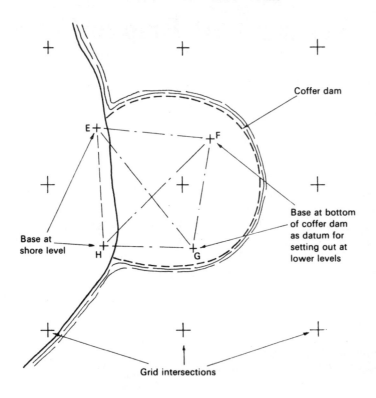

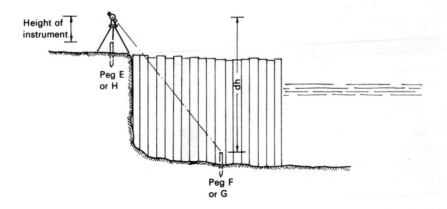

Fig. 10.2 Marine works − sub-waterline setting out.

11 Setting Out Bridge Works

The two key factors in setting out bridge works are the centre line position and the width and location of the gap. The centre line is usually easy to establish as its ends can be located from accessible positions and detail. The gap is not always easy to measure and may be water-filled or used by traffic of some sort. The choice of measurement to the accuracy required, which is usually quite high, lies in one of the following methods:

(1) Calculation from small bases on either side (braced quadrilateral figure)
(2) EDM.

Keeping to the principle of the double check, any distance for a gap measured by this method can be checked by a triangle or braced quadrilateral observed with a check base measured by ordinary linear measure. The centre line should first be established on either side of the gap and checked. Two points near to the gap should now be set out carefully upon it; they should be referenced so that once the measurement is done it is not lost. The gap between these two points can now be measured by the method chosen and the value at once used as a check of the chainage of the two points as deduced from their original settings. This also checks the centre point of the bridge itself, which should be shown as a chainage on the drawings.

If there is a discrepancy (it is not likely to be large), then a decision must be made to use one of the points as datum, altering its chainage figure if necessary, so that the most accurately measured gap distance controls the position of all associated foundations, piers, etc., without the introduction of further error.

The centre line must now be transferred down to the lower construction levels at which the foundation work is to be done, and a method devised, according to circumstance, to locate the points on it in terms of chainage derived from the selected datum. With dry gaps this may not be too difficult, but it is likely to be tiresome to achieve with wet gaps. If coffer dams have been constructed it may be still more difficult owing to obstructions to lines of sight and measurement.

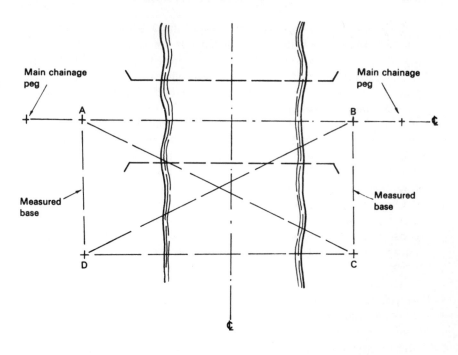

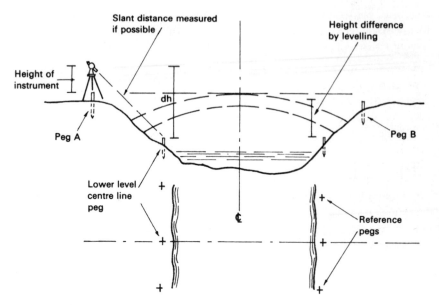

Fig. 11.1 Bridge works – gap measure and centre lines.

Slant measurement with a steel tape direct from the trunnion axis of the theodolite may suffice if the distance is not great (less than 30 m). The levels of the instrument and the lower mark to be obtained by levelling and the vertical angle should be carefully observed. When measuring from the trunnion axis the telescope should be aligned on the lower mark.

The triangle formed by difference height, theodolite peg and the slant distance measured now allows the base to be deduced in two ways as a check. From this the chainage of the lower mark is fixed in relation to that of the instrument station. Marks so established should be referenced at once. In bridging conditions some ingenuity may be required to find reference mark positions which will not be disturbed.

Once these basic marks have been fixed, all work at high and low level can now be set out with reference to centre line and chainage. As construction proceeds it becomes progressively easier, since additional datums are provided by the checked positions and dimensions of the work.

The foregoing applies to run-of-the-mill bridge works where meticulous special survey may not have been done or considered necessary. On large works a number of survey marks will have been permanently fixed, from which setting out can proceed with confidence. The key factors of centre line in both directions, over and under, still apply.

On any bridge, however, marks must be established which are secure from disturbance by construction work and from which position and level can be checked as the work goes on. This applies in either steel or concrete construction. In the latter there is the important aspect of formwork checks which must be done by the engineer both before and during construction.

Index